# REVIEWS OF UNITED KINGDOM STATISTICAL SOURCES

## VOLUME XVIII

# POSTS

AND

# TELECOMMUNICATIONS

# REVIEWS OF UNITED KINGDOM STATISTICAL SOURCES

*Editor:* W. F. Maunder

*Assistant Editor:* M. C. Fleming

REVIEWS OF UNITED KINGDOM STATISTICAL SOURCES

Edited by W. F. Maunder

*Professor Emeritus of Economic and Social Statistics, University of Exeter*

Assisted by M. C. Fleming

*Reader in Economics, Loughborough University*

VOLUME XVIII

# POSTS

AND

# TELECOMMUNICATIONS

by

STUART WALL

*Principal Lecturer and Head of Economics, Cambridgeshire College of Arts and Technology*

and

PAUL NICHOLSON

*British Telecommunications Plc*

*Published for*
The Royal Statistical Society and
Economic and Social Research Council

*by*

PERGAMON PRESS

OXFORD · NEW YORK · TORONTO · SYDNEY · FRANKFURT

| U.K. | Pergamon Press Ltd., Headington Hill Hall, Oxford OX3 0BW, England |
| U.S.A. | Pergamon Press Inc., Maxwell House, Fairview Park, Elmsford, New York 10523, U.S.A. |
| CANADA | Pergamon Press Canada Ltd., Suite 104, 150 Consumers Road, Willowdale, Ontario M2J 1P9, Canada |
| AUSTRALIA | Pergamon Press (Aust.) Pty. Ltd., P.O. Box 544, Potts Point, N.S.W. 2011, Australia |
| FEDERAL REPUBLIC OF GERMANY | Pergamon Press GmbH, Hammerweg 6, D-6242 Kronberg, Federal Republic of Germany |
| JAPAN | Pergamon Press Ltd., 8th Floor, Matsuoka Central Building, 1-7-1 Nishishinjuku, Shinjuku-ku, Tokyo 160, Japan |
| BRAZIL | Pergamon Editora Ltda., Rua Eça de Queiros, 346, CEP 04011, São Paulo, Brazil |
| PEOPLE'S REPUBLIC OF CHINA | Pergamon Press, Qianmen Hotel, Beijing, People's Republic of China |

First edition 1986

**Library of Congress Cataloging in Publication Data**

Wall, Stuart, 1946–
Posts and telecommunications.
(Reviews of United Kingdom statistical sources; v. 18)
Bibliography: p.
Includes index.
1. Postal service — Great Britain — Statistical services.
2. Telecommunication — Great Britain — Statistical services.
I. Nicholson, Paul II. Royal Statistical Society (Great Britain)
III. Economic and Social Research Council (Great Britain)
IV. Title V. Series.
HE6935.W34     1986     380.3'0941     85-29816

**British Library Cataloguing in Publication Data**

Wall, Stuart
Posts and telecommunications. — (Reviews of United Kingdom statistical sources; v. 18)
1. Postal service — Great Britain — Statistics —
Information sources 2. Postal service — Great Britain — Statistics —
Bibliography 3. Telecommunication — Great Britain —
Statistics — Information sources 4. Telecommunication —
Great Britain — Statistics — Bibliography
I. Title II. Nicholson, Paul III. Royal Statistical Society
IV. Economic and Social Research Council V. Series
380.3'07     HE6939.S7
ISBN 0-08-033967-0

*Printed in Great Britain by A. Wheaton & Co. Ltd., Exeter*

# VOLUME CONTENTS

# FOREWORD

*The Sources and Nature of the Statistics of the United Kingdom,* produced under the auspices of the Royal Statistical Society and edited by Maurice Kendall,filled a notable gap on the library shelves when it made its appearance in the early post-war years.Through a series of critical reviews by many of the foremost national experts, it constituted a valuable contemporary guide to statisticians working in many fields as well as a bench-mark to which historians of the development of Statistics in this country are likely to return again and again. The Social Science Research Council* and the Society were both delighted when Professor Maunder came forward with the proposal that a revised version should be produced, indicating as well his willingness to take on the onerous task of editor. The two bodies were more than happy to act as co-sponsors of the project and to help in its planning through a joint steering committee. The result, we are confident, will be judged a worthy successor to the previous volumes by the very much larger 'statistics public' that has come into being in the intervening years.

Mrs SUZANNE REEVE
*Secretary*
*Economic and Social Research Council*

Mrs E.J.SNELL
*Honorary Secretary*
*Royal Statistical Society*

* SSRC is now the Economic and Social Research Council (ESRC).

# MEMBERSHIP OF THE JOINT STEERING COMMITTEE

(September 1985)

*Chairman:* Miss S. V. Cunliffe

*Representing the Royal Statistical Society:*

Mr M. C. Fessey

Dr S. Rosenbaum

Mrs E. J. Snell

*Representing the Economic and Social Research Council:*

Professor J. P. Burman

Mr I. Maclean

Miss J. Morris

*Secretary:* Mr D. E. Allen

# INTRODUCTION TO VOLUME XVIII

The publication of this review has been unfortunately delayed (although the translation of British Telecom to private industry gives it coincidental topicality) through the failure to complete what was intended to have been its companion on the statistics of the road system. This would have effectively completed the coverage of the whole field of transport and communications. Earlier published work has dealt with Road Passenger transport (Review no.12), Road Goods transport (Review no.13), Ports and Inland Waterways (Review no.17), Civil Aviation (Review no.18), Rail transport (Review no.24) and Sea transport (Review no.25). In addition, there is substantial treatment of road construction and stock in Volume XII,*Construction and the Related Professions* (Review no.22). The omitted material wich was designed to form the review on the road system sources consists of a detailed coverage of stock, construction, improvements and maintenance to-gether with sources on ancillary services such as parking, traffic control, and aid patrols.

In present circumstances it is impossible to plan for the replacement of this contribution as the ESRC which has so faithfully supported the series over a long period has indicated that a term must be fixed to its committment. Hence, unless some other sponsor comes forward to assume the role, the series must end when the present contents of the pipeline are exhausted. Although reviews in progress still represent a substantial programme, it means inevitably that the project will have failed to achieve the aim with which it started of providing a complete coverage of sources in the whole economic and social field. Not only have several other reviews fallen by the wayside (which is scarcely surprising in that the main incentive for contributors is the warm glow of satisfaction that comes from the fulfillment of such a necessary professional task) but topics in other areas have not yet been commissioned while several already published now require updating. The project has accummulated a wealth of experience in its task and while it is a going concern it would need only a relatively modest grant to keep it in business but to start *de novo* or even to revive it after a lapse would be both more difficult and more expensive.

The primary aim of the series is to act as a work of reference to the sources of statistical material of all kinds, both official and unofficial. It seeks to enable the user to discover what data are available on the subject in which he is interested, from where they may be obtained, and what the limitations are to their use. Data are regarded as available not only if published in the normal printed format but also if they are likely to be released to a *bona fide* enquirer in any other form, such as duplicated documents, computer print-out, magnetic tape or disks. On the other hand, no reference is made to material which, even if it is known to exist, is not accessible to the general run of potential users. The distinction, of course, is not clear-cut and mention of a source is not to be regarded as a guarantee that data will be released; in particular cases it may very well be a matter for negotiation. The latter caution applies with particular force to the question of obtaining computer print-outs of custom specified tabulations. Where original records are held in a computer databank it might appear that there should be no insuperable problem, apart from confidentiality, in obtaining any feasible analysis at a cost; in practice, it may well turn out that there are capacity restraints which override any simple cost calculation. Thus, what is requested might make demands on computer and programming resources to such an extent that the routine work of the agency concerned would be intolerably affected.

The intention is that the source for each topic should be reviewed in detail, and the brief supplied to authors has called for comprehensive coverage at the level of 'national interest'. This term does not denote any necessary restriction to statistics collected on a national basis (still less, of course, to national aggregates) but it means that sources of a purely local character, without wider interest in either content or methodology, are excluded. Indeed, the mere task of identifying all material of this latter kind is an impossibility. The interpretation of the brief has obviously involved discretion and it is up to the users of these reviews to say what unreasonable gaps become apparent to them. They are cordially invited to do so by communicating with me.

To facilitate the use of the series as a work of reference, certain features have been incorporated which warrant a word or two of explanation. First, the text of each review is designed, in so far as varying subject matter permits, to follow a standard form of arrangement so that users may expect a similar pattern to be followed throughout the series. The starting point is a brief summary of the activity concerned and its organisation, in order to give a clear background understanding of how data are collected, what is being measured, the stage at which measurements are made, what the reporting units are, the channels through which returns are routed and where they are processed. As a further part of this introductory material, there is a discussion of the specific problems of definition and measurement to which the topic gives rise. The core sections on available sources which follow are arranged at the author's discretion – by origin, by subject subdivision, or by type of data; there is too much heterogeneity between topics to permit any imposition of complete uniformity on all authors. The final section is devoted to a discussion of general shortcomings and possibly desirable improvements. In case a contrary expectation should be aroused, it should be said that authors have not been asked to produce a comprehensive plan for the reform of statistical reporting in the whole of their field. However, a review of existing sources is a natural opportunity to make some suggestions for future policy on the collection and publication of statistics within the scope concerned and authors have been encouraged to take full advantage of it.

Secondly, detailed factual information about statistical series and other data is given in a Quick Reference List (QRL). The exact nature of the entries is best seen by glancing at the list and accordingly they are not described here. Again, the ordering is not prescribed except that entries are not classified by publication source since it is presumed that it is this which is unknown to the reader. In general, the routine type of information which is given in the QRL is not repeated verbally in the text; the former, however, serves as a search route to the latter in that a reference (by section number) is shown against a QRL entry when there is a related discussion in the text.

Third, a subject index to each review acts as a more or less conventional line of enquiry on textual references; however, it is a computerised system and, for an individual review, the only peculiarity which it introduces is the possibility of easily permuting entries.

The object at this level is merely to facilitate search by giving as many variants as possible. In addition, it also makes possible selective searches by keyword over any combination of reviews and a printout of the entries found may then be prepared. Anybody wishing to use this facility should apply to the Exeter editorial office.

Fourth, each review contains two listings of publications. The QRL Key gives full details of the publications shown as sources and text references to them are made in the form [QRL serial number]; this list is confined essentially to data publications. The other listing is a general bibliography of works discussing wider aspects; text references in this case are made in the form [B serial number].

Finally, an attempt is made to reproduce the more important returns or forms used in data collection so that it may be seen what tabulations it is possible to make as well as helping to clarify the basis of those actually available. Unfortunately, there are severe practical limitations on the number of such forms that it is possible to append to a review and authors perforce have to be highly selective.

If all or any of these features succeed in their intention of increasing the value of the series in its basic function as a work of reference it will be gratifying; the extent to which the purpose is achieved, however, will be difficult to assess without 'feedback' from the readership. Users, therefore, will be rendering an essential service if they will send me a note of specific instances where, in consulting a review, they have failed to find the information sought.

As editor, I must express my very grateful thanks to all the members of the Joint Steering Committee of the Royal Statistical Society and the Economic and Social Research Council. It would be unfair to saddle them with any responsibility for shortcomings in execution but they have directed the overall strategy with as admirable a mixture of guidance and forbearance as any editor of such a series could desire. Especial thanks are due to the Secretary of the Committee who is an unfailing source of help even when sorely pressed by the more urgent demands of his other offices.

The author joins me in thanking all those who gave up their time to attend the seminar held to discuss the first draft of his review and which contributed materially to improving the final version.

We are most grateful to Pergamon Press Ltd. for their continued support and in particular to the Production Department who put all the pieces together. The subject index entries have been compiled by Mrs. Marian Guest who has also acted as editorial assistant throughout. Special thanks are due to Mr.Ray Burnley who has masterminded our use of the Lasercomp System at Oxford University Computer Service and to the latter for the use of this facility.

W.F. Maunder

*University of Exeter*

# 31: POSTS AND TELECOMMUNICATIONS

Stuart Wall

and

Paul Nicholson

1

# REFERENCE DATE OF SOURCES REVIEWED

This review represents the position, broadly speaking, as it obtained at September 1984, with minor revisions up to the proof-reading stage of March 1985.

# ACKNOWLEDGEMENTS

The authors have recieved assistance and advice from a large number of persons and official bodies. Particular thanks are due to Scott Simpson and James Berry of the Post Office, Ian McIntyre of British Telecom, Alison Holding of the Central Statistical Office, and Fred Chambers, Librarian at the Cambridgeshire College of Arts and Technology.

# LIST OF ABBREVIATIONS

| | |
|---|---|
| BABT | British Approvals Board for Telecommunications |
| BB | Blue Book |
| BT | British Telecom |
| BTRA | British Telecom Report and Accounts |
| COE | Census of Employment |
| COP | Census of Population |
| CSO | Central Statistical Office |
| DPE | Data Processing Executive |
| EFL | External Finance Limit |
| FES | Family Expenditure Survey |
| FSBR | Financial Statement and Budget Report |
| GHS | General Household Survey |
| JICNARS | Joint Industry Committee for National Readership Surveys |
| LIS | Letter Information Survey |
| MCS | Mail Classification Survey |
| MLH | Minimum List Heading |
| NDPS | National Data Processing Service |
| NEDO | National Economic Development Office |
| NES | New Earnings Survey |
| NRS | National Readership Survey |
| PBX | Private Branch Exchange |
| PO | Post Office |
| POSSF | Post Office Staff Superannuation Fund |
| POUNC | Post Office Users' National Council |
| PSS | Packet Switched Service |
| RT | Regional Trends |
| SCNI | Select Committee on Nationalised Industries |
| SIC | Standard Industrial Classification |
| TS | Telecommunications Statistics |

# CONTENTS OF REVIEW 31

CHAPTER 1

# INTRODUCTION

This review describes the publicly available sources of statistical information about postal and telecommunications services in the United Kingdom. Under the heading of postal services it covers public letter mail and public parcel mail services and, since 1932, postal orders. Under the heading of telecommunications services it covers public telephone, telegraph, telex, data and facsimile services. Radio and television broadcasting are not included.

## 1.1 Historical Background

In the past, United Kingdom postal and telecommunications services were dominated by the Post Office (PO). Since its establishment by Charles I in 1635 the Post Office has operated public letter mail services. In 1869 the Post Office was also granted exclusive rights to operate inland telegraph services and in 1880 this monopoly was extended to cover telephone services (although private telephone companies continued to operate under licence until 1912).

The Post Office operated as a government department until 1969 when it became a public corporation, on a basis similar to that of other nationalised industries. At that stage, it was organised as four almost independent businesses. The Postal business provided letter and parcel mail services and operated the agency services to other organisations; the National Girobank offered payment and personal banking services; the Telecommunications business operated the telephone, telegraph, telex and other telecommunications services, and the National Data Processing Service (NDPS) provided computing services to the Post Office and also to outside organisations. In 1977 the NDPS was absorbed into the Telecommunications business as its Data Processing Executive (DPE).

In 1981 the giant Post Office corporation was split into two. One of the two new public corporations retained the name of the Post Office but now only has responsibility for postal, giro-banking and counter services, i.e. the previous Postal business and National Girobank. The other corporation was called British Telecom (BT) and took over responsibility for telecommunications services, i.e. the previous Telecommunications business (including DPE). Further legislation converted British Telecom from a nationalised industry into a private enterprise. Shares were issued, with 51% sold to private investors, and the British Government retaining a 49% holding. A new Office of Telecommunications was created to license and control the activities of BT and its competitors.

## 1.2 Postal Services

### 1.2.1 *The Post Office*
The Post Office is an enormous organisation, with an annual turnover in excess of £2,700 million in 1983. It handles over 10,000 million letters and 195 million parcels a year and operates a nationwide chain of around 22,000 Post Offices. It employs 176,000 people directly and 21,000 more as agents in sub Post Offices. In addition to postal services the Post Office counters also handle a wide range of agency activities including pension and social security payments, car and television licensing, and the issuing of visitors' passports. The National Girobank also offers payment and personal banking services across Post Office counters.

### 1.2.2 *The Postal Monopoly*
The 1969 Post Office Act transferred to the Post Office corporation the previous government monopoly privileges. This gave to the Post Office the exclusive right to operate a public letter service. All other Post Office activities, including parcel mail, agency work and banking, are subject to open competition.

### 1.2.3 *Other Postal Service Providers*
The parcel mail service has a number of competitors including other nationalised industries such as British Rail and National Carriers, and private enterprises such as Securicor and Security Express. In addition, a number of motorcycle messenger services are operating, especially in London.

## 1.3 Telecommunications Services

### 1.3.1 *British Telecom*
British Telecom is even larger than the Post Office. Its annual turnover is in excess of £5,700 million and it employs over 245,000 people. It operates mostof the UK telecommunications system, which handles 21,000 million calls a year from 29 million telephones. In addition to the telephone network, BT also operates the Telex network and the public data network PSS (Packet Switched Service). BT also provides, for sale or rental, a wide range of customer products, including telephones, answering machines, data modems, telex machines and private telephone exchanges. BT's Data Processing Executive provides computer services to BT and to outside organisations. In recent years BT has moved into computer bureau services such as videotex (Prestel) and electronic mailbox (Telecom Gold). BT has its own factories, which traditionally concentrated on the repair and reconditioning of old equipment, although now there is also a significant proportion of new manufacture.

### 1.3.2 *The Telecommunications Monopoly*
The 1969 Post Office Act transferred to what was then Post Office Telecommunications, the previous government monopoly privileges. These included

the exclusive right to operate public telecommunications services in the UK, i.e. telephone, telegraph, telex and data services. The sole exception in 1969 was the City of Kingston-upon-Hull where the local authority operated the local telecommunications network under licence to the Post Office. In 1972 the Channel Islands telecommunications systems were also transferred to the local governments under similar terms.

In the area of customer products, limited competition was also allowed by the Post Office. Competitors were allowed to sell direct to the public certain Post Office Approved terminal equipments. Such approval was only given for classes of products which the Post Office was either unwilling or unable to supply itself. Consequently, the concession only applied to a limited range of items, including large private telephone exchanges, telephone answering machines, some types of data modems, and facsimile machines.

The 1981 British Telecommunications Act transferred many of the monopoly privileges to the new corporation. However, it also conferred on the Secretary of State for Industry, wide ranging powers to vary BT's monopoly and to licence competitors. This enabling legislation has since been used in a number of ways. For instance, Mercury Communications Ltd has been licensed to create an independent telecommunications network to provide private network services and to interconnect with the BT network. In fact any customer terminal equipment (telephones, answering machines, modems, etc) may now be attached to the public telecommunications networks provided that it passes the basic safety and performance tests specified by the independent British Approvals Board for Telecommunications (BABT). Modern private exchanges, of any size, may now be sold, installed and maintained by suppliers other than BT, again subject only to BABT approval. It is very likely that BT's monopoly privileges will be further reduced in the future.

### 1.3.3 *Other Telecommunications Service Providers*
In the areas of telecommunications network services the city of Kingston-upon-Hull and the Channel Islands are licensed to operate local networks. Mercury Communications Ltd was granted a licence to offer private voice and data networks nationwide and started operations in 1983. In the area of customer products there are a large number of suppliers selling direct to customers, including many of the world's larger electrical and electronics manufacturers. For example, Racal and Case sell data transmission equipment, and ITT sells teleprinters. Many firms, large and small, sell telephones and answering machines.

### 1.4 Scope of this Review

### 1.4.1 *Standard Industrial Classification*
The 1968 Standard Industrial Classification (SIC) presented postal and telecommunications services under several Minimum List Headings (MLH). The most important Heading, MLH 708 (Postal Services and Telecommunications), covered:

"All Post Office establishments, except the factories manufacturing and repairing telephone and telegraph apparatus (classified in Heading 363) and the Post Office

Savings Department and Post Office Giro (classified in Heading 861); cable and radio services (excluding broadcasting and radio relay services) and other telephone or telegraph services."

This definition excluded the Post Office's manufacturing and banking activities. However, by default it included the computer service activities of the Post Office Data Processing Executive and the many agency services conducted across Post Office counters on behalf of government departments. Arguably these activities are not appropriate to the heading of 'Postal Services and Telecommunications'.

The 1968 Standard Industrial Classification has been replaced by a new 1980 *SIC* [B.11] which separately designates Postal Services and Telecommunications. However, the earlier classification has also been presented; first, because it is still in use in a number of series; second, because it may still be appropriate for time series analysis of the combined activities. Under the 1980 SIC, Postal Services and Telecommunications appear as 'Class' 79, 'Group' 790, with separate 'activity' headings:

*7901 Postal Services*

Post Office establishments primarily engaged in transmitting letters, packets, and parcels by post and providing agency services, including Crown Post Offices and sub-postmasters if their shops are wholly devoted to Post Office business. Postal service peripheral services, unless they are in separate establishments, are included but National Girobank is classified to heading 8140/2. The Post Office Philatelic Bureau is classified to heading 6480.

*7902 Telecommunications*

Telecommunications services of all kinds including Post Office telecommunications, the City of Hull Telephone Services and offices of international cable companies. Associated relay stations, both for internal and overseas transmissions, are included whether for telecommunications or for broadcasting signals. Broadcasting transmitters and local cable relay systems for radio or television are classified to heading 9741/1. The Post Office data processing service is classified to heading 8394, along with the provision of other computer service data enterprise competitors. Post Office telecommunications equipment factories are classified to heading 3441.

National Girobank is to be fully covered in a forthcoming volume in this series (*Financial Statistics* by Kate Phylaktis). Parcel delivery services run by organisations other than the Post Office - for example British Rail, National Carriers and Securicor - are now classified under heading 7230 (previously under MLH 703), and were covered by D L Munby and A H Watson in an earlier volume [B.6].

The provision, by organisations other than the Post Office, of terminal equipment for attachment to the Post Office telephone network is now classified under heading 3441:

"Manufacture of line telegraph and telephone apparatus of all kinds, including exchange equipment, switchboards, teleprinters, telewriters, etc."

In the past, such private provision of equipment has been limited to large private telephone exchanges, radio-telephone systems, telephone answering machines, facsimile machines and data transmission models. In the new environment of open competition in the supply of terminal equipment both the range of attachments and the volume of sales are likely to increase greatly. This review concentrates primarily on 'activities' 7901 and 7902, although some parts of 7230, 3441 and 8394 have also been included to give a more balanced view of those postal and telecommunications services for which the Post Office and British Telecom have private enterprise competitors.

### 1.4.2 *Geographical Coverage*

This review covers all of the United Kingdom, i.e. England, Wales, Scotland and Northern Ireland, as well as the Isle of Man and the Channel Islands. The latter two, though outside the United Kingdom, have appeared in the statistics of the service providers. Government statistics normally cover the whole of the United Kingdom. However, Post Office and British Telecom statistics usually relate to the United Kingdom of Great Britain and Northern Ireland, including the Isle of Man but excluding Kingston-upon-Hull. Prior to 1972 the Channel Islands were also included in Post Office statistics but they have been excluded since 1972.

Post Office and British Telecom regional statistics are normally based on their own administrative regions rather than on political boundaries. This leads to a number of discrepancies between PO/BT and government statistics. The most obvious example of this problem is the border between England and Wales. Government statistics use the well-defined political boundary, whereas PO/BT statistics for Wales usually covers the 'Wales and the Marches' administrative region, which includes most of Cheshire, Shropshire, Herefordshire and Worcestershire.

### 1.4.3 *Time Period Coverage*

This review concentrates on the time period from the formation of the Post Office Corporation in 1969 upto September 1984. Moreover, many of the regular sources go further back than 1969 or have predecessor publications.Hazlewood's review [B.3] gives descriptions of some of this earlier material.

### 1.4.4 *Unpublished Information*

This review mainly covers publicly available information sources. However, Government departments and service providers have a great deal of information which they do not publish but which can be made available on request. In addition to references made to unpublished data in the text, Appendix I contains a list of contact points through which further enquiries can be directed. It should be borne in mind that there are only limited resources for handling such requests in most of the organisations listed. Furthermore, service providers may have to refuse to release information they regard as commercially sensitive.

CHAPTER 2

# MAIN SOURCES

The sources of statistical information relating to postal services can be classified into three broad categories; Surveys and Reports by Service Providers, General Surveys and Reports, and other Surveys and Reports specific to postal services. Within these categories our main objective will be to identify and describe those sources which are responsible for the collection and/or publication of original information and data.

## 2.1 Surveys and Reports by Service Providers - Postal Services

**2.1.1** *Post Office Report and Accounts (PORA) [QRL.27]*
The Post Office Act 1969 requires the production, every year, of a report and a statement of accounts. This document covers financial years ending 31 March and usually appears in the following July. The 1979/80 Report and Accounts was unusual in that it covered the transition year between the reorganisation of the Post Office into 2 separate operations and the formal authorisation of this change by Parliament. Consequently it was published in 2 parts, both being extracts from the version required by the 1969 Act, Since then, and pursuant to the British Telecommunications Act 1981, *PORA* has only covered Posts and National Girobank. *Report and Accounts* was preceded between 1961 and 1969 by Report and Commercial Accounts and prior to this by Commercial Accounts. Again Hazlewood's review [B.3] gives descriptions of some of the earlier material. Although the presentation varied over time the content of these predecessors remained much the same.

*Report and Accounts* comprises 3 major sections; the Report, the Accounts and the Supplementary Statements. The Accounts and the Supplementary Statements are provided both in aggregate and also separately for each of the main businesses. From 1969 until 1977 there were 4 main businesses; Posts, Telecommunications, National Girobank and the Data Processing Service. In 1978 the Data Processing Service was merged into the Telecommunication business and, as previously noted, in 1981 Telecommunications ceased to be a part of the Post Office.

**2.1.1.1** *The Report* includes a statement from the Chairman of the Post Office and chapters reviewing the financial performance and significant developments in both Posts and National Girobank. Our concern here is the Report on Posts, which includes many useful statistics that are highly condensed versions of figures available elsewhere in *PORA*, presented in simple grahical or tabular form. Included here are statistics on Financial Performance and on obligations to Government, Operational Data on the efficiency of the Royal Mail, and data on Customer Services, Technology and

Personnel. The Report on Posts concludes with a selection of Performance Indicators, using graphs to relate achievement to target, in the current and previous four years. In the most recent edition of *PORA* the Performance Indicators include:

Achievement of Financial Target

Contributions to Government through the External Financing Limit

Quality of Service: 1st class letters (% delivered next working day)

Quality of Service: 2nd class letters (% delivered the third working day after collection)

Post Office Counter Business

Number of Post Offices

Mails operating hours

Mails man hour Productivity

Evaluated Traffic

Mails operations Hours

Real Costs per Unit Output

Postman Turnover

Energy Consumption per $M^2$ of Floor Area

Unfortunately the information presented in the Report has varied both in content and in format from year to year, so that it is not always possible to catalogue standard entries.

**2.1.1.2** *The Accounts* include the auditor's report, accounts for the Post Office as a whole, notes on the accounts and Post Office accounting practices, and separate accounts for each of the main businesses. The accounts are prepared in accordance with Section 42 of the Post Office Act 1969 and with the directions made under that section from time to time by the Secretary of State, with the approval of the Treasury. Some details of Post Office accounting practices are given each year in the chapters of Report and Accounts entitled "Post Office accounting policies and general notes" and "Notes on the accounts". The "Notes on the accounts" contain some statistical information which is not available from other Post Office sources. For example, they give details of outstanding loans from the Secretary of State and outstanding foreign loans. They also give the numbers of Post Office board members and employees in certain salary ranges.

In the latest edition of *PORA*, the separate accounts for Posts include:

Current Cost profit and loss account

Current Cost balance sheet

Current Cost source and application of funds

In each case data are available for both the present and the previous year.

**2.1.1.3** *Supplementary statements* are provided separately for each of the main businesses. These statements contain much more detail than is presented in the accounts (whose format and content are dictated by the 1969 Act). For the Posts business, the supplementary statements include;

*Value Added Statement*

*Financial Results by Services:*

Inland Mails Service - letter services, parcels, registration

Overseas Mails Service

Postal Orders
Agency Services
Services to National Girobank

*Ten Year Financial Summary* (using historical cost data, supplemented by current
cost data since 1981)
Profit and Loss Account
Balance Sheet
Flow of Funds
Retail Price Index

*Ten Year Operational Summary*
Mail Services
Motor Transport
Counter Services
Personnel

*Operational Statistics*
Mails
Business Volume (inland; overseas)
Man Hours
System Size
Counter Services
Business Volume (Postal Services, National Girobank, Pensions and
Allowances, National Savings Services, British Telecom Services, Other
Agency Services)
Man Hours
System Size

### 2.1.2 *Quality of the Inland Letter Service [QRL.33]*

Following public criticism of the letter mail service the Postal Headquarters of the Post
Office now publishes a leaflet called Quality of the Inland Letter Service. It first
appeared in 1980 and is available quarterly from the enquiry points at Head Post
Offices. The statistics are derived from a programme of continuous sampling of letters
posted and delivered throughout the United Kingdom. 145,000 letters a month are
inspected and their dates of posting and delivery are recorded. The sample size
represents about 0.02% of total postings and the statistical error limits are estimated to
be ±0.5% for the national averages. The sampling technique is described more fully in
POUNC *Report* No 17 [QRL.36].

The only statistics presented are the percentages of first class letters delivered by the
first working day following collection and of second class letters delivered by the third
working day following collection are presented for the current quarter, the same
quarter the year before, and for the 12 months ending with the current quarter. The
survey results are also presented disaggregated by Post Office Region.

**2.1.3** *The Mail Classification Survey (MCS) - now Letter Information System (LIS) [QRL.18]*

The Post Office has conducted the Mail Classification Survey (MCS) on an intermittent basis since 1975, though its replacement, the Letter Information System (LIS), is a continuous sampling system. The object of the MCS was to obtain information about the characteristics of letter mail for the purpose of marketing and operations management. The postcode forms the primary sampling unit, with every address in the UK having a postcode. For the purposes of the survey postcodes can be divided into three types:

i.   small users - one postcode covers a number of residual or small business addresses

ii.  large users - one postcode covers one address receiving 200 or more items per day

iii. very large users - as large users but receiving more than 1000 items per day

Each month a random sample of postcodes is selected from each of these groups and a day and delivery is assigned on which the survey will take place. On the appointed day, letters for the selected postcode are inspected. If necessary, sub-sampling is applied to reduce the number of items to 25. In the sorting office, just before delivery, details are noted from the exterior of the letter including:

i.   time, date and place of posting

ii.  time, date and place of delivery

iii. postage class, amount and method of payment

iv.  method and accuracy of addressing

For about 1 in 7 of the selected postcodes customer questionnaires are attached to each of the selected letters. This requests information from the interior of the letter including:

i.   nature of contents, eg greetings cards, other social, invoice, payment

ii.  type of sender. eg private, government, utility, bank, shop, customer, supplier.

The results from the MCS are not published on a regular basis although extracts have appeared in a number of places, eg Michael Corby's book, *The Postal Business* [B.1]. The LIS has now replaced the MCS, and combines quality of service monitoring with the information previously obtained through the MCS. Information derived from LIS on contents and origin of mail is outlined in CSO *Social Trends* 14 [B.10]. Any further enquiries should be addressed to the post office contact point in Appendix I.

**2.1.4** *Post Office Guide [QRL.25]*

The *Post Office Guide* is an annual publication, providing a summary of principal services and charges for Inland Post, Overseas Post, National Girobank, Postal Order Services and Savings and other services. Supplements to each current edition, giving particulars of amendments, are published approximately every two months. Since 1980 there has been a separate *British Telecom Guide* [QRL.2].

**2.1.5** *Post Offices in the UK [QRL.28]*

*Post Offices in the UK* contains the postal addresses of every post office in the UK, except London.

**2.1.6** *Postal Addresses and Index to Postcode Directories [QRL.29]*
This publication contains the correct postal addresses of approximately 25,000 place names in the UK (excluding London) and the Irish Republic which are essential in a postal address. The title of the appropriate Postcode Directory is given against each entry. Postcode Directories are available for inspection at major Post Offices.

**2.1.7** *London Post Offices and Streets [QRL.17]*
This provides the detail for London omitted from the two previous publications.

## 2.2 Surveys and Reports by Service Providers - Telecommunications

**2.2.1** *British Telecom Report and Accounts (BTRA) [QRL.3]*
First published in 1982, and annually since then, *BTRA* is the main source for financial data on British Telecom, although some operational data are also presented. *BTRA* is split into two broad sections;

**2.2.1.1** *The Report*
This contains a table giving financial results for turnover and profit, with the latter prepared under the historical cost convention modified by the charging of supplementary depreciation to reflect replacement cost. Financial performance is also evaluated, relating achievement to target, and giving details of the part of the external financing limit utilised, together with interest paid on government loans and the change in real unit costs. The Chairman's Statement and a breakdown of the enterprises included in BT, is then followed by a breakdown of operational statistics (including quality of service) for Inland Services, ie Telephones, Telex and Data Transmission. A further section of the Report presents operational statistics for British Telecom International, again including quality of service statistics. The Report concludes with a section on major systems, providing details of Research and Development activity and expenditure in the previous year.

**2.2.1.2** *The Accounts*

The Auditor's Report is followed by an outline of Accounting policies. The Profit and Loss Account, Statement of Movement in Reserves, Balance Sheet and Source and Application of Funds are presented in separate tables. Notes to the account give further breakdowns of selected items. A Ten Year financial summary gives data for profit and turnover, the components of the self-financing ratio, capital employed, financial return, tariff indices and a breakdown of changes in annual turnover and costs by cause (price change/ pay/ business expansion). In the supplementary statements to the accounts, the annual financial results are disaggregated by service type, distinguishing Inland from International Services. Annual operating costs, staff numbers and capital expenditure are also disaggregated in separate tables. The Accounts conclude with a selection of current cost statements.

**2.2.2** *Telecommunications Statistics (TS) [QRL.42]*
Every year, usually in June, the Telecommunications Headquarters of the Post Office publishes a booklet called *Telecommunications Statistics*, and commonly known as 'the orange book'. It is the main source for operational data and first appeared in 1933, containing a wealth of detail about Post Office telecommunications activities. The figures presented are mainly compiled from Post Office operational records although in some cases these are supplemented by estimates. Historical statistics are available for every year but for brevity the published tables are limited to every fifth year prior to 1970. Figures for intervening years can be obtained on request. As far as possible these time series are presented in a consistent manner. In a few cases, where it has been necessary to change the basis of a particular statistic, the change is clearly marked both in the tables and in the footnotes.

*Telecommunications Statistics* contains information about telephone traffic, exchange connections, telephone exchanges, line plant, telephone traffic, telegraphs, telex, datel, manpower, motor transport, finance, tariffs and telecommunications history. Most of the information presented is abstracted from records kept for operational reasons; for example customer records, equipment records, and staff records. However, some statistics have to be estimated. One example of this relates to figures about the number of local telephone calls made. Brian Smith describes the methodology of this estimation technique in his dissertation [B.7].

The following provides a rather more informative picture of the statistics available from *TS*.

*Notes*
*Definitions*
*Telephones*
*Stations*
Annual Totals and Net Annual Increase
Annual Totals by Rental Classification (business; residence; call office; BT and
    service; private circuit; radiophones)
Annual Totals by Type of Exchange (manual; automanual; automatic-strowgeer/
    crossbar/ electronic)
*Densities* (connexions per 100 population; stations per 100 population; telex
    connexions per 1000 business connexions)
*Exchange Connexions*
Annual Totals and Net Annual Increase
Demand and Supply
Annual Totals by Rental Classification (as above)
Annual Totals Shared Service (business; residence)
Public Telephone Call Offices (kiosks-urban/ rural; cabinets; open call offices; total
    receipts; average receipts per call office)
Annual Totals by Type of Exchange (as above)
*Equipment and Line Plant*
Exchanges by Type and Size
Number of Exchanges of Each Type
Exchange Plant - Total and Spare Capacity
Customers' Cables, Cross Connexion and Distribution Points
Trunk Lines

*Traffic*
Number of Originated Effective Calls (inland-local/ trunk; international-transatlantic)
Analysis of International Calls
Growth of Originated Effective Calls
Annual Calling Rate
Calls to Information Services (Discline; FT Cityline; Traveline; Recipeline; Bedtime stories; Weatherline; Leisureline; Cricketline; Raceline, Sportline; Localine)
Analysis of Inland Trunk Calls
*Telegraphs*
*Equipment*
VF Systems and Channels
*Traffic*
Inland Service
International Service
*Telex*
Traffic
Exchanges and Connexions
*Datel*
Datel Services (total modems)
*Manpower*
Deployment and Functional Disposition
*Motor Transport* (Vehicles; trailers; vehicular mechanical aids, workshops and fuel; fleet totals and fleet mileages)
*Finance*
Expenditure and Income
Fixed Assets
*Historical*
Telephone Tariffs and Services
Telegraph Tariffs and Services
Telex Tariffs and Services
Telecommunications - Growth and Development

### 2.2.3 *Quality of Telephone Services [QRL.32]*

Following criticism of the quality of the public telephone service the Telecommunications Headquarters of the Post Office now produces a booklet called *Quality of Telephone Services*, commonly known as 'the green book'. It first appeared for the financial year ending 31 March 1979 and is now published annually, usually in June. Quarterly updates are available on request.

British Telecom monitors a small random sample of all telephone calls to determine whether they fail or succeed, with the reasons for failure noted. The primary purpose of these observations is to check on the adequacy of equipment provision and maintenance, and to identify which parts of the network need improvement. Similarly, calls to the operator are sampled to check the adequacy of switchboard manning levels and to identify problem areas. The results of these surveys are presented as measures of the quality of the automatic telephone service and the operator service. However, the statistics have to be interpreted with care because of the limitations of this technique. In particular the observation points chosen tend to bias the results in favour of British Telecom (see for example 3.6.2 below). Fuller detail of the survey methodology is given in the green book itself.

BT records all faults reported to it together with the time of the report and the time of fault clearance. The statistics of the quality of the repair service are derived from these records. Two statistics are presented; the average number of faults per telephone per year, and the average percentage of faults cleared by the end of the next working day.

### 2.2.4 *A Report to Customers [QRL.37]*
Published annually, and based on the British Telecom Reports and Accounts for that year, *A Report to Customers* presents a useful synthesis of financial and operational data.

### 2.2.5 *List of Exchanges [QRL.16]*
Every year BT publishes a booklet called *List of Exchanges*. This first appeared in 1913 and adopted its current form in 1922. It lists all the telephone and telex exchanges in the system, and for each exchange gives the type of equipment installed, and the numbers of connexions and stations served by the exchange. All figures are derived from Post Office exchange plant records.

### 2.2.6 *City of Kingston-upon-Hull Telephone Department [QRL.7]*
The local telephone service in the city of Kingston-upon-Hull is operated by the city corporation under licence from the Post Office. Although the Telephone Department has not published any statistics since its 1970 brochure, updates on the 1970 figures are available from their Public Relations Officer. The information available is limited to spot figures for the size of the area served, its estimated population, the numbers of exchanges, telephones, connexions, public call offices and rented coin offices, telephones per 100 population, and the number of staff employed. Capacity figures (numbers of lines) are also available for every exchange in the Hull area.

### 2.2.7 *States of Jersey Telecommunications Board - Annual Report [QRL.40]*
Every year since 1973 the States of Jersey Telecommunications Board has published an annual report. This contains conventional accounts including a trading and profit and loss account, a balance sheet, and a source and application of funds statement. Notes on the accounts include figures for the amount borrowed from the States Capital Fund, and the cost, depreciation and net book value of fixed assets. It should be noted that the Jersey financial year runs from 1 January to 31 December which differs from the year used by the Post Office. The *Annual Report* also gives some non-financial information. For example the most recent issue gives staff numbers, local cable capacities, numbers of exchange lines and telephone instruments installed, motor vehicle statistics and number of telephone calls.

**2.3 General Surveys and Reports - Postal Services and Telecommunications**

Here we highlight a number of sources which, though presenting statistical material relevant to the UK as a whole, are nonethelesss original sources for collection and/or publication of material relating to the Postal and Telecommunications industries.

**2.3.1** *Census of Employment*
Until June 1971, apart from the Census of Population (see below), most employment data derived from a count of National Insurance cards. However, following the announcement in 1969 that National Insurance cards were to be discontinued, a new system was devised (see Department of *Employment Gazette*, August 1973, for a comparison of the two systems). Census forms were now to be despatched each June to pay points, ie offices from which employers send their PAYE statements to the Inland Revenue. The pay point is asked to show the numbers of employees for whom it holds pay records. Separate figures are sought for males and females and for full and part time employees, and a brief description of business activity is required. Pay points are to provide such data for each address for which they hold pay records. Analyses can then be presented by sex, industry and locality, distinguishing full and part time employees. Of interest to us is that for MLH 708 'Postal Services and Telecommunications', annual figures are available for number of employees, by region and sex, separating full time from part time employees.

A full Census is conducted every third year, with 1982 such a year. For other years, forms are not sent to employers with less than 3 employees at the last full Census. Although there are a large number of such employers they only account for some 1.5% of all employees. The Census of Employment is published in *Employment Gazette* [QRL.8] and *British Labour Statistics Yearbook* [QRL.1].

**2.3.2** *Census of Population (COP) [QRL.4]*
With the exception of 1941, full censuses of population have been held decennially for England and Wales, and for Scotland, since 1801. The first such census for Ireland was taken in 1821, then decennially until 1911. Thereafter, following the Government of Ireland Act of 1920, censuses were taken in Northern Ireland in 1926, 1937 and at ten year periods since 1951. These censuses provide an original source for data on the previous MLH 708 and now for group 790 'Postal Services and Telecommunications'. For example, in the 1971 Census MLH 708 'Postal Services and Telecommunications' is disaggregated by:
   a.   Status and sex of employees. GB; England and Wales; Scotland
   b.   Male employees by age/females by age and marital status. GB; England and Wales; Scotland.
   c.   Employees by sex and area of workplace. GB; England and Wales; Scotland.
   d.   Employees by occupational type and by sex. GB.
   e.   Males by hours worked, females by hours worked and marital status. GB; England and Wales; Scotland.

f.     Employees by Social Class, Socio-economic class, and sex. GB; England and Wales; Scotland.

In the 1981 Census, as well as data for group 790, the following tables show activity within group 790 separately, ie 7901 Postal Services and 7902 Telecommunications.

a.     Usually resident population aged 16 and over in employment; Industry Division and Classes by Employment Status by Sex. GB; Scotland.

b.     Usually resident population aged 16 and over in employment; Occupation Orders and Groups by Industry Classes by Sex. GB; Scotland.

In the 1961, 1971 and 1981 Censuses, data as regards 'Postal Services and Telecommunications' are available only on a 10% sampling of census returns. The sampling method adopted, together with estimates of sampling errors, are presented in the Census Reports themselves.

### 2.3.3 *Census of Production [QRL.5]*

A Census of Production was first conducted in the UK in 1907, then periodically until 1948, when annual censuses of a summary character were instituted, interspersed with full, detailed censuses. Since 1970 a new system of annual censuses had been introduced, covering all manufacturing establishments in the UK with 20 or more employees, accounting for approximately 95% of the total turnover of the UK manufacturing and mining industries.

A number of censuses have provided detail of annual expenditure by each Minimum List Heading (MLH) industry on the item 'postage, telephone, telegrams, cables and telex'. Such data are available for 1948, 1963 and 1968 for establishments employing over 300 persons. Unfortunately this item cannot be disaggregated into constituent components, and no further detail of any kind is available other than that published.

### 2.3.4 *Family Expenditure Survey (FES) [QRL.9]*

The *Family Expenditure Survey* (FES) (1953/4, 1957 annually) is well documented as to the methodology employed [B.4], providing detail on incomes and patterns of expenditure for different groups of households in the UK. A set sample of 11,000 addresses is selected annually. As well as information via interview, each household maintains a detailed expenditure record for 14 consecutive days. The sample is designed so that each household has an equal chance of selection, and so that the interviews are spread evenly throughout the year.

Average weekly household expenditure on the item 'postage', telephone, telegrams' is recorded for a variety of household types. For example, in the latest report this information is available by households with differing ranges of income; by composition (one adult retired/non-retired households, one man-one woman retired/non-retired households, by zero, one, two, three or more children) and income of household; by type of Administrative Area (Greater London, Metropolitan districts, non-Metropolitan districts, by high/low population density); and by region. Data are also provided as to the numerical distribution of telephones by household income, by household composition, and by tenure of dwelling.

It is of particular interest that, via unpublished data, the item 'postage, telephone, telegrams' can, on request to the Department of Employment, be disaggregated into three separate component items as follows:

a.  Postage, including parcel post, poundage on postal orders and money orders.
b.  Telephone Account.
c.  Telephone, other than account; telegrams, cables.

These components are available for all years since 1961 and for all the household types for which an expenditure table appears in the *FES* report of that year. Apart from the household types already mentioned, this means that in the most recent report these components are also available by occupation and age of the head of the household, by tenure of household and for households with and without car(s). As well as absolute expenditure on the item 'postage, telephone, telegrams', or its components, we can readily derive for each household type the *percentage* of total expenditure allocated to that item or component. Approximate standard errors of means and proportions are available on request to the Department for each item or component, together with the number of recording households. The table content of the *FES* is revised each year so that expenditure for a household type included in the latest report may not be available for all years.

### 2.3.5 *Financial Statement and Budget Report (FSBR) [QRL.10]*

For the financial year 1975/6, and annually thereafter, the capital reqiuirements of the nationalised industries are directly related to their sources of finance. For our purposes it is of interest that the 'Post Office' is separately classified, with the following breakdown:

*Capital Requirements*; Fixed Assets in the UK, other assets.

*Sources of Finance*; Internal resources, Government grants, Government issues of public dividend capital, net borrowing from the National Loans Fund, net overseas borrowing, other domestic borrowing, leasing.

Using estimated prices for each year, data are provided for the past financial year and a forecast made for the forthcoming year. Although the raw data for the items above are furnished to the Treasury by the Post Office itself, the Financial Statement and Budget Report does constitute an original source as regards data presentation, if not collection. In the first place, the publications of the Service Providers (such as *PORA* prior to 1981, and now *BTRA*, and *TS*) do not provide such detail in relating capital requirements to the sources of finance. Secondly, data presented in FSBR is tailored so as to conform with the principles and methods used in National Income Accounts Classification (see [B.5]). Finally, no forecasts are available from publications of service providers.

### 2.3.6 *General Household Survey (GHS) [QRL.11]*

Most of the data presented in *GHS* have already been made available via *FES* above. There is however one quite different table in *GHS*, which annually gives the percentage of households with telephone by socio-economic group of the head of household. The groups designated include professional; employers and managers; intermediate non-manual; junior non-manual; skilled manual and own account non-professional; semi-skilled manual and personal services; and unskilled manual. There is also an annual breakdown of telephone possession by economically inactive heads (retired/other).

**2.3.7** *Input - Output Tables for the UK [QRL.14]*
MLH 708 'Postal Services and Telecommunications' features in the input-output tables
compiled for the UK. Tables based on full information were compiled for 1954, 1963,
1968 and 1974.Annual tables, based on partial information used to update the 1968
tables, are available from 1970 to 1973. These tables do *not* constitute a primary data
source, since in most cases the sources of data for MLH 708 are the Service Providers
themselves (see above) and Censuses of production (see above). However, the
input-output tables are most certainly a primary publication source, presenting tables
which reveal in a unique way, interdependencies between 'Postal Services and
Telecommunications' and the rest of the economy.

A complete description of methodology employed in constructing the UK
input-output tables is given in the 1968 volume, which distinguishes 90 industry and
commodity groups. The tables for 1970 also feature 90 such groups, although those for
1971 and 1972 only present 59 groups, and that for 1973, 35 groups. A full account of
the methods used to derive the annual tables is presented in *Economic Trends* (April
1975, June 1978 issues), with the RAS procedure [B.12] playing a prominent role. In
fact, it was recent doubt as to the accuracy of RAS in updating absorption matrices as
the period from the base year lengthens (see *Economic Trends*, April 1975), that
resulted in the 1973 tables containing only 35 separate groups. However, though less
accurate, copies of the 1973 tables in 59 group detail can be obtained from the Central
Statistical Office on request. The 1974 full information census is as firmly based as that
for 1968 and distinguishes 102 industry and commodity groups. In all the input-output
tables, MLH 708, 'Postal Services and Telecommunications' appears as a separate
group, though it is often designated 'Communications' in the tables themselves.

The input-output tables provide a wealth of information as regards Postal Services
and Telecommunications:
1.   A commodity analysis of intermediate purchases from domestic production, by
     industry group.
2.   Final demand components: Current Expenditure (consumers; public
     authorities); Gross Domestic Capital Formation (Fixed; Stocks); Exports of
     goods and services.
3.   Total requirements per 1000 units of final industrial output/domestic
     commodity output, in terms of both gross and net output.
4.   Industrial output in terms of primary input, in co-efficient form.
5.   Industrial composition of final expenditure in terms of net output, in
     co-efficient form.
6.   Direct and indirect import and tax content of industrial output, in co-efficient
     form.
7.   Total and Direct requirements of imported commodities per 1000 units of
     domestic output.
8.   Analysis of public authorities current expenditure (military defence; National
     Health Service; other central government; local authorities).
9.   Plant and machinery investment matrix.
Each set of input-output tables is constrained to the national accounting aggregates
in the National Income and Expenditure Blue Book for a specific year. It should also
be noted that the Central Statistical Office has re-run the 1963, 1968 and 1970
input-output tables to give a more comparable industrial classification for these years.

These are published in Background Paper 6 of the National Economic Development Office, 'Study of UK Nationalised Industries'.

**2.3.8** *National Income and Expenditure Blue Book (BB) [QRL.19]*
BB provides a primary publication source as regards data for final expenditure on the output of MLH 708. The source of data is mainly the service providers themselves, often in response to specific requests from the CSO for supplementary data to that presented in *Post Office Report and Accounts* and *Telecommunication Statistics*. BB is useful in presenting final expenditure data in conformity to strict National Income Accounting conventions. It should also be noted that data are occasionally presented under a broad group heading of 'Transport and Communications'. In most cases data can be provided on request which relates solely to 'communications', a pseudonym for MLH 708 itself.

**2.3.9** *National Readership Survey (NRS) [QRL.20]*
These annual surveys provide a primary data source for characteristics of households owning telephones. In particular, ownership is classified by possession of a variety of consumer durables, by readership of a wide range of publications, and by region. The surveys are administered by JICNARS (Joint Industry Committee for National Readership Surveys). The sample is designed to produce 30,000 interviews in a full year, with the basis of the sampling method being a pre-selected sample drawn from the National Register of Electors. Data from the Census of Population is used to produce a sampling frame of local authority areas stratified by the Registrar General's planning regions, by area type, socio-economic index, etc. A full description of sample design and interview questions is provided in each published survey.

**2.3.10** *New Earnings Survey (NES) [QRL.21]*
The New Earnings Survey (1968, 1970 annually GB, 1971 annually Northern Ireland), is based on a 1% sample of employees according to National Insurance numbers, for one specific pay period in April. The results relate to men aged 21 and over on 1 January of the year in question, and women aged 18 and over on that date, who work in excess of 30 hours per week (ie 'full-time' employees), and whose pay for the reference period was not affected by absence. Volume XIII of the present series, by A.Dean [B.2] gives a full description of the NES. For our purposes we should note that for MLH 708, 'Postal Services and Telecommunications', the following data are provided:

  a.   Average gross weekly earnings, hourly earnings and weekly hours. Full-time manual and non-manual men and full-time manual women.
  b.   Annual increases in average gross weekly and hourly earnings. Full-time manual and non-manual, men and women. Based on matched samples. This information is given both as an absolute value and as a percentage increase.
  c.   Distributions of gross weekly and hourly earnings. Full-time manual and non-manual, men and women. Percentages 'less than' specified amounts, and various quantiles.

d.   Make-up of gross weekly earnings. Full-time manual and non-manual, men and women. Overtime, piecework, shift payments separately classified.

In all the above cases, standard errors, where applicable, are presented in the tables of the NES. Results are only *published* for sample units in excess of 100 persons, and when standard errors are within 2 percentage points of the mean. Although the results for MLH 708 normally fulfil these criteria, whenever this is not so, unpublished data are usually available on request to the Department of Employment for sample units in excess of 50 persons, having standard errors within 4 percentage points of the mean.

### 2.3.11 *Regional Trends (RT) [QRL.34]*

This presents an annual breakdown of the percentage of households with telephone by region. Although not a primary data source it is a primary publication source.

### 2.3.12 *Social Trends (ST) [QRL.39]*

Although the source of the various statistics presented in ST is British Telecom and the Post Office, it does present a useful summary of the main trends in communication activities. In particular ST provides a useful ratio on the relative price of telephone calls and mail over the past ten years. The telephone call used in the ratio is that of a cheap rate, direct dialled call, over 56 km in distance, and lasting 3 minutes (including VAT). The item of mail used in the ratio is a first class letter weighing up to 60 grams. The ratio has been quite volatile since the early 1970s, when the price of the telephone call was around two-thirds higher than that of the letter. During the 1970s the ratio fell progressively, so that by 1979 the price of the telephone call was only around half the price of the letter. Since then the ratio has risen again and the telephone call in 1983 is again higher in price than the letter, by about one fifth. ST also points out the available data for assessing the growth of teletext services, which use the telephone or broadcasting systems to link TV sets to computer data banks. There has been a rapid growth in these services, including Ceefax of the BBC, Oracle of ITV, and Prestel of BT. In fact in July 1983 the one millionth teletext set was sold in the UK.

### 2.3.13 *'Strikes in Britain' Manpower Paper No 15 C.T.B.Smith, R.Clifton, P.Makeham, S.Creigh and R.V.Burn [QRL.41]*

This is the report of an internal research project of the Department of Employment official series of statistics on industrial stoppages. Stoppages recorded are those relating to terms and conditions of employment, which involve more than 10 workers and which last longer than one day. The statistics only reflect stoppages of workers directly and indirectly involved (out of work, though not a party to the dispute) at the establishments where the disputes occurred. More information as to definitions and qualifications of the data used is given in the Department of Employment *Gazette* 86 No 6 July 1977 pp 579-586).

For MLH 708 'Postal Services and Telecommunications' (as for all other MLH groups) the following data are presented:

a.   Number of stoppages per 100,000 employees, by MLH. Annual figures, 1970-75 UK.

b.   Number of working days lost per 100,000 employees, by MLH. Annual figures 1970-75 UK.

Both the above tables can be traced back to 1966 via data presented in the February 1976 issue of the Department of Employment *Gazette*. Prior to 1966 data are only available on stoppages by 50 industry groups, with no separate classification for 'Postal Services and Telecommunications'.

### 2.3.14 *United Kingdom Balance of Payments [QRL.44]*

The UK Balance of Payments records the absolute annual value of invisible trade credits and debits in 'Telecommunications and Postal Services'. This aggregate comprises overseas receipts and expenditure arising from international telephone, telegraph and telex services, the UK share in Satellite Systems, Submarine cable work and Surface mail. The figures are based both on returns by the Post Office and on Exchange Control Records. Quarterly estimates consistent with the annual figures are published in the September issue of *Economic Trends*. In fact, more detailed component statistics of credits and debits are available on request to the Head of the Balance of Payments Branch at the CSO. Quarterly data are separately available as far back as 1973 for:

Overseas telegraph
Telex
Telephone
International leased circuits
Surface mail, parcel post and transit
Air Mails
Submarine cables and satellite projects

Half yearly information back to 1968 is also available for these components separately, and back to 1958 for these components combined.

### 2.3.15 *Which? Reports [QRL.45]*

Consumer Association 'Which' reports (1957 - monthly) are a useful source for *ad hoc* information on quality of postal and telecommunications services, as well as comparative price data across countries. This was particularly helpful when such data did not appear in *PORA*. Each separate survey has its own design, and reference should be made to the appropriate 'Which' publication. For telephone services see July 1966, December 1968, September 1969, April 1970, February 1976, June 1976 and February 1979. For postal services see March 1969, April 1972, May 1973, November 1974, June 1976, August 1976 and June 1979.

## 2.4 Other Surveys and Reports specific to Post and Telecommunications

### 2.4.1 *House of Commons Committees*

The Select Committee on Nationalised Industries (SCNI) is empowered by the House of Commons to examine the conduct of the Nationalised Industries. A number of Sub-Committees cover a range of industries, and have authority to examine ministers, government officials and nationalised industry management. Two reports on the Post Office have originated from the SCNI. The 1967 report *The Post Office* [QRL.22] was the first comprehensive enquiry into the Post Office by an outside body for over 20 years. This is certainly a primary publication source, with the Appendices containing a wealth of information on postal/telecommunication operations and performance, much of it unavailable from regular publications by the service providers.

A second report by SCNI, *The Post Office's Letter Services*, was published in 1975 [QRL.26]. Although the Service Providers are again the source of data, a range of supplementary information is presented which is not readily available elsewhere. For instance, trends in letter traffic in the UK (Ireland and overseas) are compared with 11 other countries 1964-1973; an international comparison is also provided across EEC countries for printed papers postal rates. Detailed data are presented for staff hours in mails operations by type of activity.

An Industry and Trade Committee has been established under a new committee structure within the House of Commons. Its first report, in session 1979-80, was on the Post Office [QRL.23]. Again, supplementary evidence submitted to this committee constitutes a primary publication source. For example, useful tables were provided permitting international comparisons in delivery services and postage rates between the UK and a number of other countries; also, breakdowns of types of transactions within sub offices, and data indicating the effect on sub-postmasters' incomes of reductions in unit totals caused by loss of counter business.

The fifth report of the Industry and Trade Committee in the session 1981-82 was also on the Post Office [QRL.24] following shortly after its separation from British Telecom. This report updates the tables in the first report, and includes other primary data, such as a detailed breakdown of cost allocations for letter mail services.

### 2.4.2 *Report of the Post Office Review Committee, Chairman C.F. Carter [QRL.35]*

The Carter Committee in seeking to build up an appropriate data base commissioned a special study of postal rates and costs by H. Rees and A. Baker of the Public Sector Economic Research Centre, Leicester University. Tables presented include:

i.   1914-1971 selected periods, index of rates for ordinary inland letters (1914 = 100) as compared with the retail price index (1914 = 100).
ii.  1912-1976 selected periods, time series of surplus/deficit on postal services.
iii. For 1922-1939 (1922 = 100) and 1965-1976 (1965 = 100), indices of mail volume handled per worker, as compared with a mail volume index.
iv.  1965-1976 (1965 = 100) selected periods, index of administrative labour cost per mail unit, as compared with an index of mails operations labour cost per mail unit.

### 2.4.3 *The Beesley Report on the Liberalisation of the use of the British Telecommunications Network [QRL.15]*

The Beesley report in 1981 considered the economic implications of allowing freedom to offer services to third parties over the BT network. In the course of the study a number of useful tables are presented, not easily accessed from other sources. These include:

A cross-country comparison of tariffs for 100 km telephone circuits.

Measures of data transmission penetration by country, including a forecast by Logica.

A review of price elasticity co-efficients in various UK Inland Trunk Call Studies.

Net price elasticity in the US for customer dialled Station to Station calls by mileage band.

A review of Price Elasticity Coefficients in various International Telephone Demand Studies.

**2.4.4** *The Monopolies and Mergers Commission*
In September 1979 the Secretary of State referred the operations of the London Letter Post to the Monopolies and Mergers Commission. The subsequent report in March 1980 [QRL.13] has provided extremely useful data on quality of service and productivity performance, not only in London but also in the provinces when used as a comparator. Particularly useful is a time series, 1970-79, relating national trends in delivery performance for first and second class mail to those observed in London. The latter were disaggregated into performance statistics for letters posted in the London Postal Region and delivered elsewhere in the UK, and those posted in the UK and delivered in London. A full review of sampling methods adopted is available in POUNC *Report* No 17 [QRL.36]. Original data are also provided by the Post Office for productivity trends 1967-1979, separately designating the Inner London Area, Outer London Area and Other Regions. Productivity is defined as Evaluated Traffic divided by Adjusted Gross Paid for Hours. Evaluated Traffic is itself defined as Letters posted plus 2.5 × Parcels Handled (see 3.5 below). Adjusted Gross Hours are Gross Paid for Hours, Minus Hours attributable to Transit Traffic etc.

**2.4.5** *The Post Office Users' National Council (POUNC)*
POUNC was established by the Post Office Act 1969, as a consumers' watchdog with members appointed by the Government. To date over 30 reports have been published, one for every package of tariff proposals and service cuts, together with reports on specific issues such as quality of service. These reports constitute in part a primary source, as with *Report* No 4 [QRL.12] which presented a survey of delivery of first class mail, to be used as a check on delivery figures published by the Service Provider itself in *PORA*. Similarly *Report* No 23 [QRL.30] provides primary data on delivery of 1st and 2nd class mail, as well as on delivery performance for letter packets, rebate mail and parcels. *Report* No 28 [QRL.31] investigated the Postal Service overseas mail performance, again presenting primary data. Still more important is the fact that the various reports constitute a primary publication source for a wide variety of operational data on postal services.

CHAPTER 3

# AVAILABLE DATA

The most comprehensive statement of available data, by type, is presented in the Quick Reference List. The intention here is rather that of making some qualitative statements on the data that are available.

## 3.1 Labour Statistics

With some 411,000 employees, the Post Office, prior to its separation in 1981, was the largest commercial employer in Europe, providing almost 2% of total UK employment. Now BT employs 246,000 persons and the Post Office some 178,000 persons. However, in relation to the value of capital employed, Postal Services are the more labour intensive in operation (see 3.2 below).

### 3.1.1 *Employment*
Annual totals of employment are provided in *BTRA* [QRL.3] and *TS* [QRL.42] for BT, and in *PORA* [QRL.27] for the Post Office. For annual detail on employment in the whole of MLH 708 'Postal Services and Telecommunications', including private sector counterparts, by region and sex, distinguishing full and part-time employees, reference can be made to COE [QRL.1] and [QRL.8]. Decennial data are available from COP [QRL.4] for 'Postal and Telecommunication' employees by industrial status and socio-economic class; and for age distribution by MLH industry. In the 1981 census postal services and telecommunications are separately identified (see 2.3.2 above) using the new 1980 SIC classification.

### 3.1.2 *Unemployment*
Quarterly data for numbers unemployed by industry and sex are presented in *British Labour Statistics Yearbook* [QRL.1], with MLH 708 'Postal Services and Telecommunications' separately represented.

### 3.1.3 *Occupational Type*
Little detail is available on occupational type from *PORA*, *BTRA* or *TS*. There is some annual description of the 'nature' of employment, but this does not accord with conventional occupational classification (see [B.8]). For example, from *PORA*, employment in Posts is enumerated under the following headings: Post Offices, mail

31

operations, engineering, motor transport, Postal Headquarters, Regional Headquarters, Supplies Department, Philatelic Bureau, Postal Order Service and National Television Licence Records Office. The breakdown for Telecommunications employment from *BTRA* is General Management, Staff Administration, Training, Catering and Office Services, Finance and Management Services, Planning and Works, Service and Marketing, Research and Development, Purchasing and Supply, Factories, Accommodation, Motor Transport and Data Processing. Detailed decennial data for MLH 708 by occupational type, and sex, can be obtained from COP [QRL.4], and since 1981 for Posts and Telecommunications separately, ie 'activities' 7901 and 79012 respectively. Such data are extremely useful for comparison across other SIC industries, though its infrequency, and periodic revisions of occupational title, make inter-temporal comparison more difficult.

### 3.1.4 *Wage Rates and Earnings*
*PORA* provides annual data for total employee compensation, broken down into pay, pensions, employers' social security contributions, and sub-postmasters' renumerations. *BTRA* also provides annual data for total employee compensations, with the only disaggregation available being the number of employees in various salary ranges over £20,000, together with emoluments paid to the small number of members of the corporation. Annual data for MLH 708 'Postal Services and Telecommunications' is available from *NES* [QRL.21], which gives median; quartiles and deciles of gross weekly/hourly earnings, by sex and MLH industry. Also designated are percentages of employees with earnings 'less than' specified amounts.

Wage rates for specified manual occupations are presented annually, as at 1 April, in *Time Rates of Wages and Hours of Work* [QRL.43]. Postmen (by grade), Telegraphists, Telephonists, Postal Officers, are separately designated under 'Manipulative Grades'. Labourers, Technicians (by class and length of service), Technical Officers, are presented under 'Engineering Grades'. In addition to basic time rates of wages, normal weekly hours of work for which time rates are payable are given. Also supplements, guaranteed payments, rates for overtime, for 'young' workers (less than 20 years), and paid holiday entitlement. Changes in wages and conditions are reported at intervals in the monthly supplement to the *Employment Gazette*, 'Changes in Rates of Wages and Hours of Work' [QRL.6].

Wage rates for the occupational category 'postmen, mail sorters and messengers' are separately identified in *NES*, which gives gross weekly average earnings (with/without overtime), gross hourly earnings (with/without overtime), and pay (£ per week) at various percentiles of the income distribution for this occupational category.

## 3.2 Current Inputs other than Labour

The cost of labour input includes, say, pensions and employers' share of Social Security contributions. Before the split in 1981 such labour costs comprised over 59 per cent of total expenditure for the Post Office as a whole. However, whereas labour input formed over 78 per cent of postal service expenditure, the corresponding amount for telecommunications services was only 47 per cent. It is in the latter that inputs other than labour are of greatest significance.

### 3.2.1 *Apparatus and Equipment*

*PORA* provides data for annual expenditure by Posts on various types of apparatus and equipment within the total of fixed assets. The data are limited to totals for net additions to plant and machinery, vehicles, computers and other equipment, as well as land and buildings. Data as regards the efficacy of the Mechanised Letter Office programme is not in fact available from regular publications. One must turn instead to irregular sources, the most recent of which is the *Report of the Post Office Review Committee* [QRL.35]. Estimates of past and future costs and savings, in annual present value equivalents, are provided both by the Post Office and by independent assessors.

Data on apparatus and equipment are more disaggregated for telecommunication services with *BTRA* giving separate totals for additions to, and net expenditure on, inland transmission equipment (duct, cable, radio and repeater equipment), exchange equipment (strowger, crossbar, semi-electronic, digital), customer equipment, computers, office machines and furniture, motor vehicles, and miscellaneous equipment. Similar data are also provided for international cables and equipment, and for net expenditure on materials awaiting installation. Until 1981 *TS* also provided a detailed analysis of Fixed Assets at Net Book values, with plant and accommodation finely disaggregated, along with motor transport, office machines, materials awaiting installation and communications satellite. Unfortunately, this series has been discontinued. *TS* provides further volume detail for apparatus and equipment within telecommunications services, such as the number of exchanges by type (eg Strowger, Crossbar, Electronic, etc) and size (number of exchange connexions), and subscribers' cables and telephone trunk lines by type. The *Input-Output tables* of 1968 [B.12] and 1974 [QRL.14] present a plant and machinery investment matrix, indicating purchases by postal services and telecommunications (MLH 708) across 29 and 39 commodity groupings respectively.

### 3.2.2 *Raw materials and other intermediate domestic inputs*

Little information is available from the service providers in either *PORA*, *BTRA* or *TS* as regards intermediate purchases absorbed in the provision of final output. The most useful sources are undoubtedly the input-output tables for the UK [QRL.14]. As has already been mentioned, these are not a primary data source, since in most cases the service providers and various censuses of production are the sources of data for MLH 708, though they are most certainly a primary publication source. Tables based on full-information were compiled for 1954, 1963, 1968 and 1974. Annual tables are available from 1970 to 1973, based on a partial updating of the 1968 tables. The detail available can be gauged from the fact that the 1974 tables indicate intermediate purchases by postal services and telecommunications from 102 domestic industry and commodity groups. NEDO Background Paper 6 [B.13] is useful in evaluating both direct and indirect (inputs into inputs) inputs of a number of industries into posts and telecommunications for 1963, 1968, 1970 and 1971.

### 3.2.3 *Imports*

A detailed commodity composition of imports in the provision of domestic output by postal services and telecommunications is not directly available from the service providers. Again the most useful sources are the UK input-output tables referred to above, in which separate tables are presented for commodity analyses of imports, as at the 102 commodity level for 1974. Useful derivative tables are also provided. Such as total requirements (direct + indirect) of imported commodities per 1000 units of domestic output.

### 3.3 Research and Development

Until 1980/81 *PORA* provided annual data for total expenditure on research and development by postal services and telecommunications separately. Since 1980/81 *PORA* no longer gives regular information on Research and Development expenditure/personnel for postal services, though *TS* does give separate figures for annual expenditure on Research and Development by British Telecom, and numbers employed in the Research and Development function. *BTRA* gives a more detailed breakdown of Research and Development expenditure by system function, networks and switching, transmission, customer systems and services, general technology, and business support. It is, of course, in telecommunication services that microelectronic and related technologies are invoking most rapid changes in capital equipment, networks and peripherals. The stress laid upon technological progress in telecommunication services can be gauged from the fact that research and development constitutes almost 2 per cent of total expenditure in the most recent year. In contrast, for postal services the corresponding figure, when last presented, was less than one twenty-fifth of one per cent of total expenditure.

### 3.4 Output

#### 3.4.1 *Posts*
The Supplementary Statement for Posts in *Post Office Report and Accounts* [QRL.27] is the primary source in this area.

**3.4.1.1** *Letters* The operational statistics table in the Supplementary Statements to *PORA* shows the number of letters posted each year, sub-divided into first and second class for inland letters and into surface rate and surcharged airmails for overseas letters. It should be noted that the separate figures for registered items, recorded delivery items and business reply items are totals for letters and parcels; and the figures for overseas correspondence at surface rate include letters carried to Europe by airmail without surcharge.

The financial results by services table in the Supplementary Statements to *PORA* gives the income and profit of the Post Office from inland letter services. The figure for income and profit from overseas mails service includes parcel traffic.

**3.4.1.2** *Postal Orders* These were transferred to Posts in 1982, with the number and value of postal orders issued, together with the financial outcome of this service, separately identified in *PORA*.

**3.4.1.3** *Parcels* The operational statistics table in the Supplementary Statements to *PORA* shows the number of parcels handled by the Post Office each year, sub-divided into inland (including Irish) and overseas. Overseas parcel traffic is further sub-divided into surface rate and surcharged airmails, and also into inward and outward traffic. The separate figures for registered and recorded delivery items include letter mail as well as parcels. The financial results by services table in *PORA* gives the income and profit of the Post Office from inland parcel services. The figure for income and profit from overseas mails services includes letter mails.

### 3.4.2 *Telecommunications*
*BTRA* [QRL.3] and *TS* [QRL.42] are the primary sources for this information.
**3.4.2.1** *Telephone calls* Statistics on the numbers of telephone calls are derived in part from call counting meters in telephone exchanges, in part from the billing records for calls handled by the operator, and in part from estimates. Estimation is only used for the number of dialled local calls.

Prior to the introduction of Subscriber Trunk Dialling (STD) each subscriber's dialled local calls were counted on his own meter at the telephone exchange. All other calls were handled by the operator. Consequently, analysis of the billing records gave accurate counts of the number of calls made. With the introduction of STD in 1958, subscribers' meters were modified to count call charge units rather than calls. In some designs of exchange other meters were available on which calls could be counted. However, in many local exchanges the ability to count local calls was lost on conversion to STD.

British Telecom estimates the total number of local calls by extrapolating from the figures produced by those exchanges having local call counters. These exchanges cover only one-third of the total exchange connections and are predominantly based in the big cities. Consequently the inherent assumption of a nationally uniform calling rate per connection may not be valid. B W Smith explains the methodology and its limitations more fully in his Diploma Thesis [B.7]. *TS* [QRL.42] presents annual totals for the number of calls disaggregated into inland calls and external (international). Inland calls are further sub-divided into local and trunk calls. External calls are sub-divided into continental (European), intercontinental and maritime (ship-to-shore). These categories are further sub-divided into outward calls (from one foreign country to another but passing through the UK), with the percentages of continental and intercontinental calls dialled also designated.

Three derived statistics are presented. Call growth rate is expressed as the percentage increase over the preceding year. Calling rate per exchange station (each telephone instrument capable of being connected with a public exchange) and per connection (each exclusive exchange line and each shared line counts as one exchange connection) are also separately specified. Annual totals for calls to recorded information services are disaggregated by service (eg Dial-a-Disc, Speaking Clock, Weather etc). Of particular interest here is the Speaking Clock which is by far the most popular, receiving over 400 million calls a year.

**3.4.2.2** *Telegrams* Telegram traffic figures were, until 1982, presented in *TS* as the number of inland and external (international) telegrams transmitted each year. Inland telegrams ceased in 1982, but until then were sub-divided into public service, press, railway and PO service groups. The number of inland telegrams are sub-divided into telegrams with foreign administrations, maritime radio telegrams and traffic with the Irish Republic, with each group further sub-divided into inward, outward and transit categories.

**3.4.2.3** *Telex* Telex traffic is presented in *TS* divided into inland and external (international) traffic. Year end totals are given for the number of inland telex calls, both manually controlled and dialled, and for the number of dialled meter units (charge units). For external traffic, year end totals are given for the numbers of manually controlled calls, for the numbers of dialled calls and the metered units incurred, and for the numbers of call minutes used (sub-divided into outward, inward and transit).

## 3.5 Productivity

### 3.5.1 *Postal Services*

For a full discussion of alternative indices for measuring productivity in postal service, see Annex 8 to the Monopolies and Mergers Commission Report on *The Inner London Letter Post* [QRL.13]. One such is the productivity index calculated for 1967-1979, separately designating Inner and Outer London areas, and other Regions, presented in the Monopolies and Mergers Report quoted above. A simplified efficiency index is used:

$$\text{Productivity} = \frac{\text{Evaluated traffic processed}}{\text{Gross paid for hours during processing}}$$

$$\text{ie Productivity} = \frac{(N1 + 2.5N2)}{(M1 + M2 + M0)}$$

$$
\begin{aligned}
\text{where } N1 &= \text{Number of letters posted} \\
N2 &= \text{Number of parcels handled} \\
M1 &= \text{Man hours processing letters} \\
M2 &= \text{Man hours processing parcels} \\
M0 &= \text{All other paid hours}
\end{aligned}
$$

The concept of 'evaluated traffic' allows the Post Office to use data collected continually in each office, ie posted letters and parcels handled. However, for consistency, 'parcels handled' must be converted to an equivalent number of 'parcels posted' and then weighted to reflect the fact that the relative work content of letters and parcels is approximately 1 to 6. In fact a multiplier of 2.5 is sufficient for this conversion because, at the point of counting, the number of parcel handlings is already 2.4 times the number of parcels posted.

This same index is now presented annually in *PORA* in the section 'Performance Indicators', under the heading Mails Man-Hour Productivity. *Ad hoc* sources also provide productivity series, as with the study by M Rees and A Baker (*op cit*) specially commissioned by the Post Office Review Committee, Chairman C F Carter. For 1922-1939 (1922 = 100), and 1965-76 (1965 = 100), indices of mail volume handled per worker are compared with a mail volume index. Again, in Table 2 of the Appendix to the report of the Post Office Review Committee, an index of hours used and business levels is available from 1965 to 1976.

An implicit index of total factor productivity is provided annually in the 'Performance Indicators' of *PORA* in the guise of changes in real unit costs. A five year graph shows the annual percentage change in the cost of producing one unit of output, adjusted for inflation as measured in the Retail Price Index. If any of these series are used as indices of productivity, it should be noted that they do not take account of changes in quality of service (see below), and therefore provide at best only a partial picture.

As regards counter services themselves, *PORA* provides annual data on volume of transactions, numbers engaged as counter staff, number of Crown Offices (run exclusively by the Post Office) and number of Sub-Post Offices (operated on an agency basis). Deriving productivity indices from such data is even more difficult. For example, the work equivalents of different types of transaction are not available; yet the issuing of, say, postal orders differs in time from the payment of pensions. Changes in the mix of the total volume of transactions cannot, therefore, be evaluated in terms of labour input. In addition, *PORA* does not permit the total volume of transactions to be allocated between Crown and Sub-Post Offices. Although the number of Sub-Post Offices has fallen relative to the total of transactions, productivity implications are unclear since variation in Sub-Post Office share of transactions is unknown.

### 3.5.2 *Telecommunication services*
Labour productivity is a concept less readily ascribed to the provision of telecommunication services. This follows from the fact that the industry is highly capital intensive, with frequent variation in the mix of technological vintage. Of course annual data are available from *TS* on percentage growth in business volume, capital formation and staff numbers, and selective indices such as telephones per employee are also available. Again an implicit index of total factor productivity is provided annually in *BTRA*. A ten year index shows the changes in volume of business (ie revenue net of price changes, showing additional income due to business expansion).

## 3.6 Quality of Service

### 3.6.1 *Postal services*
Statistics on the quality of letter mail services appear in a number of publications. However, most of these statistics are derived from the same source, the Post Office's sample survey. Since 1980 the Post Office has published its survey results quarterly in its *Quality of the Inland Letter Service* leaflet [QRL.33]. However, extracts from the results of earlier surveys have appeared in a number of enquiry reports; for example the *Report of the Post Office Review Committee* [QRL.35], the Select Committee on Nationalised Industries report *The Post Office's Letter Services* [QRL.26], the Monopolies and Mergers Commission report on *The Inner London Letter Post* [QRL.13], and the Post Office Users' National Council *Reports* Nos 17, 23 and 28.

Few independent sources are available. There have been a few small scale surveys conducted and published by newspapers and periodicals. The most reliable of these are the ones by *Which?* [QRL.45] though based on smaller sample sizes than the Post Office survey. Consequently they are less representative of the postal services although they do give more varied indices of performance. See in particular the June 1979 report for parcel services and November 1974 for letter services. The POUNC Reports mentioned above also include primary data on quality of service statistics.

### 3.6.2 *Telecommunication services*
The primary source of information in this area is the Post Office's green book, *Quality of Telephone Service Statistics* [QRL.32]. It has sections covering the automatic telephone service, the telephone operator service, and the telephone repair service. The statistics for the automatic telephone service need to be interpreted with care. Because inland trunk calls are sampled at the trunk exchanges and international calls are sampled at the international gateway exchanges, the survey results do not reflect any problems occurring before the monitoring points. Consequently the results tend to underestimate the extent of call failures due to insufficient or faulty equipment. *BTRA* [QRL.3] and British Telecom's *A Report to Customers* [QRL.37] also give summary details of the Quality of Service Statistics. All the quality of service statistics are presented as average performances. The original data are not available, nor is any indication given of the spread of the results, so that no assessment can be made of 'best' and 'worst case' performances. *Which?* reports [QRL.45] are again an independent source for quality of service statistics. For instance the February 1976 report gives some time series data on percentage failure of internal (UK) telephone calls, though the sample is rather small.

## 3.7 Prices

Postal and telecommunication service prices have a combined weight of only 0.018 in constructing the Retail Price Index. Yet changes in prices for these services are subject to intense scrutiny given their widespread use and their importance to particular sectors of the economy, such as publishing and mail order.

### 3.7.1 *Postal Services*
*PORA* provides details of tariff changes in each financial year for postal services. Also available is a ten year tariff index for mails services adjusted for inflation. Since 1977/78 *PORA* has provided a comparison of letter prices across a number of countries for 1st class delivery. Prices are for a minimum weight and converted to pence using purchasing power parity exchange rates. Each package of proposed changes in tariffs since 1969 has been the subject of a separate POUNC *Report* [QRL.38]. Unpublished data are often elicited from Service Providers during the course of these enquiries. These reports therefore constitute an extremely useful primary publication source for *ad hoc* data underlying proposed price changes. For instance, trend data are often given highlighting component changes in both sources of income and items of expenditure,

and forecasts made of income and expenditure with and without the proposed price changes. Comparative price data are frequently provided, by service type, across a range of countries.

*Which?* [QRL.45] is another useful source for *ad hoc* data on prices, though in this case it is both a primary data source (having initiated its own surveys) and primary publication source, with the June 1976 (postal and telephone services) and June 1979 (parcel services) editions containing price data. Unusual insights are available from this source, as in the June 1976 report assessing how long a phone call could be made before it became cheaper to send a letter, by type of call, distance, and type of letter. Within the Retail Price Index as a whole, a separate index is available for the component 'postage'. *Employment Gazette* [QRL.8] presents a monthly index based on January 1974 = 100 for postal prices, together with percentage changes over the previous one and twelve months respectively.

### 3.7.2 *Telecommunications Services*

*TS* provides a chronological documentation of principal changes in telephone and telegraph tariffs since 1912, and telex tariffs since 1954. Annual data are available from *BTRA* estimating the additional telecommunications income that can be adduced to price changes in that year, as opposed to other factors such as expansion of business. Also available from *BTRA* is a ten year tariff index, both unadjusted and adjusted for inflation, together with comparative data on the cost of typical telephone bills. Within the Retail Price Index as a whole a separate index is available for the component 'telephones and telemessages'. Again a monthly index (January 1974 = 100) is available in *Employment Gazette* [QRL.8] together with percentage changes over the previous one and twelve months.

## 3.8 User Statistics

It is perhaps useful to examine separately two types of user statistics. The first relates to user expenditures, for instance whether demand is intermediate or final, and if final whether for private or public consumption, capital formation, or export. The second relates to user characteristics, such as family circumstance, social class, or business type.

### 3.8.1 *User expenditures*

National Account statistics make a clear distinction between intermediate and final purchases from MLH 708, postal services and telecommunications. Intermediate purchases are, at least ostensibly, those which are not desired for their own sake but in order to be 'used up' in the course of some subsequent stage in the production process. Final purchases are, in contrast, deemed to be desired for their own sake.

**3.8.1.1** *Intermediate demand purchases by industry* A number of Censuses of Production [QRL.5] provide detail of annual expenditure by each Minimum List Heading industry on the item 'postage, telephone, telegrams, cables and telex', namely those for 1948, 1963 and 1968 for establishments employing over 300 persons. Unfortunately such disaggregation is not available from later censuses. We have already noted the usefulness of UK input-output tables [QRL.14] in providing detail of intermediate purchases by posts and telecommunications. Such tables are equally relevant here in that they provide details of intermediate sales by posts and telecommunications to other MLH industry groupings (one hundred and two such groupings separately identified in the most recent tables).

**3.8.1.2** *Final demand purchases* As we noted, final output in the National Accounts consists of items which are not used up in the production of other goods and services. Such items must be either
- i.    used up for their own sake - ie consumed, in the form of current expenditure by private individuals or by public authorities, or
- ii.   used to increase the nation's wealth, or its stock of capital, ie invested.

Investment (or Gross Domestic Capital Formation) is by convention of two types; fixed, incorporating the purchase of assets such as machines or factories; and stocks, including changes in the value of finished or semi-finished inventories in the productive pipeline. The output of MLH 708 is classified to the above categories of domestic final demand purchases. In addition, overseas final demand is recorded under the heading 'exports'. Measurements following such conventions are presented annually for MLH 708 and its SIC 1980 equivalent in the *National Income and Expenditure Blue Book* (BB) [QRL.19] and occasionally in the *Input-Output Tables* [QRL.14]. In both instances service providers are a main source for raw data, though aided by selected CSO sample surveys. However, these are most certainly primary publication sources, with raw data facing the stringent constraints of National Income Accounting conventions.

BB provides annual data at both current and constant prices for consumers' expenditure on 'postal services' and 'telephone and telegraph services' separately for the previous ten years. Figures for current expenditure by General Government on goods and services of MLH 708 are available on request. As published in BB they are grouped under the heading 'Transport and Communications' but disaggregated data are available. As regards Gross Domestic Capital Formation, data are only available for postal, telephone and telegraph services in combination. Fixed capital formation at current prices is presented for the previous ten years, in total, and broken down into vehicles, ships and aircraft, plant and machinery, and new buildings and works. MLH 708 is not separately designated as regards stocks.

BB does not provide data for MLH 708 re the final demand component 'exports'. Recourse must instead be made to *UK Balance of Payments* [QRL.44]. As was mentioned in the description of this source in Section 3, data are presented annually for receipts arising from international telephone, telegraph and telex services, the UK share in Satellite Systems, Submarine cable work and Surface Mail. Although published data only present the annual total of such receipts, the value of each item is available separately, on request, as far back as 1973 and in quarterly time periods. Half yearly information can be traced still further back for each item, to 1968, and for all items combined to1958. The presentation of data in the UK Balance of Payments is broadly consistent with National Income Accounting conventions, and where departures do occur these are specified.

### 3.8.2 *User characteristics*

A main source for data on user characteristics is the *Family Expenditure Survey* (FES) documented in Section 2.3.4 above. Annual data are available from 1957 as to average weekly household expenditure on 'postage, telephone, telegrams'. This information is available by households with differing ranges of income, by composition and income of household, by type of Administrative area, and by region. On request to the Department of Employment, data can be provided separately for postage; telephone account; and telephone other than account, telegrams and cables. In addition, FES provides data on the numerical distribution of telephones by household income, by household composition, and occasionally by region. *Regional Trends* [QRL.34] presents a regular annual breakdown of the percentage of households with telephone by region.

The *National Readership Surveys* (NRS) [QRL.20] provide a further primary data source for user characteristics. Annual data are available for telephone penetration nationally, by region, by possession of a variety of consumer durables, and by readership of a wide range of publications.

Few data are available regularly as regards user characteristics for postal services from *PORA*. *Ad hoc* surveys are conducted internally by the Post Office into mail flow characteristics, such as the volumes of letters classified by source of origin - destination, such as business - business, business - residential, residential - business and residential - residential. The *Mail Classification Survey* [QRL.18] is one such internal survey. *BTRA* provides annual data on the user source of both income and profit for telecommunication services. Income, profit (loss) and return on capital employed in Inland Telecommunication Services is allocated to Subscribers' calls, Call office receipts, Private circuits, Telegrams, Telex and Agency services. Rental income, profit and return on capital employed is allocated to Business, Residential and Apparatus.Similar, though still more detailed data on the user source of income are presented annually in *TS*.

## 3.9 Current Revenue and Expenditure of Service Providers

*Post Office Report and Accounts* [QRL.27] is the primary source for postal services, with the most useful data appearing in the Accounts and in the Supplementary Statements. The figures given in *BTRA* [QRL.3] are the primary source for telecommunication services, with extracts also presented in *TS* [QRL.42].

### 3.9.1 *Postal services*

The 'Financial Results by Services' tables of *PORA* give separate figures for turnover and profit (loss) for inland letter, parcel and overseas services, together with counter services, postal orders and other services. Value Added Statement of *PORA* presents the total expenditure of postal services disaggregated by broad function, such as inland conveyance, overseas conveyance, finance and banking, accommodation and motor transport. The Current Cost Profit and Loss Account of *PORA* sets Staff Costs, Depreciation and other operating expenditures against turnover. A still more detailed breakdown of expenditure by function for letter mail services is available in Appendix 4 of the fifth *Report* from the Industry and Trade Committee [QRL.24].

### 3.9.2 *Telecommunication services*

For the Telecommunications business separate figures are given in *BTRA* for rentals, call charges, private circuits, telegrams (until 1982), telex and international services. The 'ten year Financial Summary' tables gives an analysis of income growth, sub-dividing it into additional income due to price changes and additional income due to increased volume of business. For telecommunication services, current expenditure is broken down in the Profit and Loss Account of *BTRA* into Staff costs (wages and salaries, social security costs, other pension costs), Depreciation and other operating charges. The 'Details of Operating Costs' table in *BTRA* gives a detailed sub-division of operating expenditure into functional categories. The 'Ten Year Financial Summary' in *BTRA* presents an estimate of the increase in operational costs due to changes in pay and price levels, and that due to an increased volume of business.

## 3.10 Capital Expenditure and Depreciation

### 3.10.1 *Postal services*

Post Office capital expenditure on postal services is identifiable as the additions to the historical cost of fixed assets in the Postal business. *PORA* [QRL.27] gives this information in tabular form in the 'Report' section. The figures are presented disaggregated by major classes of asset, for example land and buildings, plant and machinery, motor vehicles, computers and other equipment. However, it is not possible to identify the proportion of these new assets used in the operation of the postal services only. At least some part of these assets relate to National Girobank and to the provision of agency services. In the 'Notes on the Financial Statements' *PORA* makes provision for accumulated depreciation, adjusted for current replacement costs disaggregated by major classes of asset.

### 3.10.2 *Telecommunications*

*BTRA* [QRL.3] is the main source of information about telecommunications capital expenditure. The 'Details of Capital Expenditure table' gives the additions to the historical cost of the fixed assets of the Telecommunications business. The figures are presented disaggregated by major classes of asset, for example land and buildings, type of inland transmission equipment, type of inland exchange equipment, customer equipment, computers and office machines, motor vehicles, and international cables and equipment. The Fixed Assets table in *BTRA* 'Notes to the Accounts' also gives the depreciation provision based on historical costs, disaggregated by major classes of asset. The total figures for depreciation based on historical cost and adjusted for current replacement costs (supplementary depreciation) are given in the Profit and Loss Account. It should be noted that the Post Office's auditors have always qualified the accounts with respect to the supplementary depreciation allowance because it is based in part on estimates of existing fixed assets.

### 3.11 Taxes, Subsidies and External Finance Limits

For many years the Post Office has ploughed its trading profits back into the business for use in its enormous capital investment programme. As a result of various capital allowances it has usually avoided liability to Corporation Tax. During 1973/4 and 1974/5 the Post Office, at government insistence, pegged its prices despite rapidly rising operational costs. This resulted in large trading losses, especially in 1974/5. In these two years the government paid a subsidy to the Post Office which appears in the accounts in the form of 'compensation for price restraint'.

After major tariff increases in 1975/6 Post Office trading profits soared and eventually exceeded the Pay Code reference level. The Post Office refunded £101 million to its customers in the form of a discount on the bills presented in 1977/8. This item appears in Reports and Accounts for 1976/7 as 'provision for elimination of profit above the pay code reference level'. The postal service has in recent years been required to contribute towards the financing of the Public Sector Borrowing Requirement. In effect this has taken the form of a negative external finance limit (EFL). For instance Posts was required, after meeting all its expenses and financing its capital investment programme, to contribute £55.8 m to general Government funds in 1982-3. BT has, in contrast, been granted a positive EFL by the Government in recent years. For instance it could borrow £380 m in 1982, though in fact it only utilised £250 m of this facility. Until 1982 borrowing had to be from the National Loans Fund, but in that year BT was given permission to raise money directly from the markets such as by issuing Telecom bonds.

### 3.12 Current Assets and Liabilities

#### 3.12.1 *Postal Services*
*PORA* gives figures for current assets and current liabilities in the Balance Sheet tables. Current assets are given disaggregated by major class of asset; stocks, debtors, investments, and cash and short term funds. Current liabilities are also disaggregated by major class; creditors, agency service balances, and advances from other businesses. One recurring feature in *PORA* is the pension fund deficiency. Although not shown in the Balance Sheets as a current liability it has been a continuing drain on Post Office finances for over a decade. Prior to the formation of the Corporation, Post Office employees benefited under civil service pension provisions. However, in 1969 the Post Office Staff Superannuation Fund (POSSF) was established in order to make staff pensions independent of the government. Actuarial valuations in 1973 and 1976 showed the Fund to have insufficient assets to meet its long term liabilities. Regular payments have been made to the Fund by the Post Office to rectify this position and these will continue until 1992.

#### 3.12.2 *Telecommunications*
*BTRA* gives figures for current assets and current liabilities in the Balance Sheet, which are further disaggregated in the Supplementary current cost statements and in notes to the Accounts. Current assets are broken down by stocks, debtors (trade; prepayments

and accrued income), short-term investments and cash at bank and in hand. Current Liabilities are broken down by short-term borrowing,trade creditors (including taxation and social security) and accruals and deferred income. A detailed analysis of loan capital is presented in notes to the Accounts.

## 3.13 Fixed Assets

### 3.13.1 *Postal Services*

*PORA* [QRL.27] is the primary source for data on fixed assets. In the 'Notes on the Financial Statements' *PORA* shows the gross replacement cost, accumulated depreciation, and net replacement cost of Postal business assets. The figures are presented disaggregated by major classes of asset - land and building, plant, furniture and office machines, motor vehicles, and assets in course of construction. There is no indication of the proportion of these assets employed in the provision of postal services rather than agency services. Very little information is available about the numbers of postal assets as distinct from their value. However, the Operational Statistics table of *PORA* does give year-end totals for the number of motor vehicles in use, for the numbers of Crown Post Offices and Sub Post Offices, the number of delivery points, and the number of fully mechanised letter and parcels offices. It should be noted that Sub Post Offices are not owned by the Post Office and therefore are not counted in their asset valuation.

### 3.13.2 *Telecommunications*

**3.13.2.1** *Value of fixed assets.* BTRA is a major source for data. The Fixed Assets table in 'Notes to the Accounts' shows the historical costs, accumulated depreciation, and net book value of Telecommunications business assets. The figures are presented disaggregated by major classes of asset - land and buildings and plant and equipment. Plant assets are further disaggregated into cables and transmission equipment; telephones and related equipment; exchange equipment; telex and other miscellaneous equipment; and international cables and equipment. *TS* [QRL.42] gives a much finer disaggregation of Fixed Assets at net book values. Annual data are presented for plant, broken down by type, for accommodation by type, motor transport, office machines, materials awaiting installation, and communication satellites.

It should be noted that the Post Office's auditors have persistently qualified the accounts with respect to the figures on fixed assets. The problem relates to the traditional Post Office practice of depreciating its assets over estimated average lifetimes rather than actual lifetimes. Consequently the figures for asset values certainly include some allowance for assets which have actually been scrapped. On the other hand some assets still in active use have been written out of the books. The Post Office (and now BT) is making steady progress towards compiling fixed asset registers to satisfy the auditors. However, the qualifications are likely to persist for at least another ten years, although the value of assets should reduce steadily.

**3.13.2.2** *Telephone Exchanges. List of Exchanges* [QRL.16] is the primary source for data on telephone exchanges. It names every exchange in the country and gives for each its function in the network (local, trunk, automanual), the type of equipment installed, the numbers of business and residential connections served, and the total number of stations served. Regional totals are also given. *TS* [QRL.42] is basically a secondary source but it presents some useful analysis of exchanges by type and by size; it also gives figures for the net (ie used) capacity and the spare capacity. It is important to note the distinction between telephone connections and telephone stations. Each exclusive exchange line and each shared line customer counts as one connection. Each telephone instrument which can be connected to a public exchange is a station. Where a customer has a private branch exchange (PBX), each telephone at an extension point is a station. There are always more stations than connections in the network.

**3.13.2.3** *Telephone Connections. List of Exchanges* [QRL.16] gives the number of business and residential connections served by individual exchanges. *Telecommunications Statistics* [QRL.42] gives some useful analyses. Totals are given disaggregated into business and residential, into shared and exclusive service, and by type of exchange. The numbers of Public Call Offices of various types are also given.

**3.13.2.4** *Telephone Stations. List of Exchanges* [QRL.16] gives the total number of stations served by individual exchanges. *TS* [QRL.42] gives some useful analyses, with totals disaggregated into business, residential, call office, Post Office service, private wire and radiophone stations, and by type of exchange. Telephone densities are also given, expressed as stations per connection and stations per hundred population.

**3.13.2.5** *Cables. TS* [QRL.42] gives some very basic information. The numbers of cable pairs in the local network (from the exchange to the customers' premises) together with the numbers of primary and secondary cross connection points and the number of distribution points. See the *Institute of Post Office Electrical Engineers' Journal* [B.9] for a description of the Post Office local line network. *TS* also gives some figures for the number of long distance (more than 40 Kilometers) trunk circuits in use, disaggregated into public service, Post Office use, and private service.

**3.13.2.6** *Telegraph Equipment. TS* [QRL.42] provides data on the numbers of multi-channel voice frequency and time division multiplexing systems and circuits in use. The circuit totals are disaggregated into public telegraph, private service, and Telex service.

**3.13.2.7** *Telex Equipment. TS* [QRL.42] again gives basic information in the form of the number of Telex exchanges (manual and automatic), the number of trunk circuits between exchanges, and the number of exchange connections (manual and automatic).

**3.13.2.8** *Datel Modems* The only information available here is the number of Post Office Datel Service modems in use, given in *TS* [QRL.42]. It should be noted that this figure does not include the large number of privately owned data modems being used on rented private circuits.

**3.13.2.9** *Motor Vehicles* Again *TS* [QRL.42] is the primary source. It gives figures for the total number of vehicles used, disaggregated by type of vehicle and also by type of use (ie engineering, supplies, passenger and telegraph). Incidentally, figures are also given for the number of motor transport workshops, the volume of fuel used and the total mileages run by each of the four fleets.


### 3.14 Sources of Finance

The Post Office and BT obtain finance from three main sources; operating profit, allowance for depreciation of assets, and loans from external sources. *Post Office Report and Accounts* [QRL.27] gives such an analysis of financial sources for postal services in the Source of Application of Funds tables. The 'Notes on the Financial Statements' gives details of outstanding loans from the Secretary of State, from Public Dividend Capital and from foreign sources. *BTRA* [QRL.3] provides similar data for BT. More disaggregated data are provided in *FSBR* [QRL.10], with the sources of finance broken down by internal resources, Government grants, Government issues of public dividend capital, net borrowing from the National Loans Fund, net overseas borrowing, other domestic borrowing and leasing. Using estimated prices for each year, data are presented both for the past financial year and a forecast made for the forthcoming year. FSBR data are further constrained to accord with the practices of National Income accounting, though the primary data source remains the Service Providers.

CHAPTER 4

# FUTURE NEEDS

One of the major needs of users of data in the posts and telecommunications area has been the separate presentation of data for each respective service. This became still more necessary with the 1981 separation of posts from telecommunications. Fortunately the new 1980 SIC includes separate 'activity' headings for postal and telecommunications services, so that government statistics are progressively providing information separately, instead of in combined form (as with the previous 1968 SIC heading of MLH 708, 'Postal Services and Telecommunications').

## 4.1 Publication of Methodologies

Post Office sources give statistics compiled by a variety of different methods. Some statistics are derived by analysis of the full records kept for internal purposes. Others are based on sampling techniques. Yet others are estimates based on combinations of counting and sampling. Generally neither the Post Office nor British Telecom publishes full methodologies, although there are a few exceptions. It would be useful if they were to do so in future, or at least indicate which figures are counts, which are samples, and which are estimates.

## 4.2 Range of Quality of Service

Quality of service statistics, both for postal and telecommunications services, only give figures for the average performance of the services. It would be much more informative if some indication were given of the range of service performance. For example, on letter mail services it would be useful to give the percentages of letters delivered on the day of collection and 1, 2, 3, 4, 4 and more days afterwards. Similarly for the telephone repair service it would be useful to know not just the average number of faults per telephone per year but also the numbers of telephones having no faults, and 1, 2, 3, or more faults in a year.

## 4.3 Restriction of Data Availability via Commercial Sensitivity

There will be an increasing tendency for telecommunication suppliers, especially BT, to become protective towards commercially sensitive information. In particular, data on the customer base, market share, and market penetration will be jealously guarded. It may therefore be helpful for the government to publish aggregate data whilst protecting individual suppliers share data, but it is by no means certain that all suppliers would co-operate with such a scheme.

# QUICK REFERENCE LIST - TABLE OF CONTENTS

# QUICK REFERENCE LIST

## QUICK REFERENCE LIST

| Type of data | Breakdown | Area | Frequency | QRL Publication | Text Reference |
|---|---|---|---|---|---|
| **Output** | | | | | |
| *Postal Services* | | | | | |
| *Letters* | | | | | |
| Number posted | Inland, by 1st and 2nd class; overseas, by Europe and rest of world. | UK | Annual | [QRL.27] | 3.4.1 |
| Inland letter traffic index | | UK | Annual | [QRL.27] | Ten year series in the Supplementary Statements |
| Overseas letter traffic index | | UK | Annual | [QRL.27] | Ten year series in the Supplementary Statements |
| Percentage of total letters postcoded | | UK | Annual | [QRL.27] | |
| Turnover and profit (loss) of inland letter services | | UK | Annual | [QRL.27] | |
| *Parcels* | | | | | |
| Number handled | Inland | UK | Annual | [QRL.27] | 3.4.1 |
| Number handled | Overseas, by outward/inward | UK | Annual | [QRL.27] | |
| Total parcels handled | | UK | Annual | [QRL.27] | |
| Parcels handled index | Inwards and overseas | UK | Annual | [QRL.27] | Six year series in the Report section Ten year series in the Supplementary Statements |
| Turnover and profit (loss) of inland parcels services | | UK | 1982 | [QRL.24] | |
| Percentage market share Letters and parcels (mails) | By carrier Number registered: inland | UK UK | 1982 Annual | [QRL.24] [QRL.27] | |
| | Number recorded: inland | UK | Annual | [QRL.27] | |

| | | | | |
|---|---|---|---|---|
| Number of business reply and freepost correspondence: inland | UK | Annual | [QRL.27] | |
| Number of registered and insured items: overseas | UK | Annual | [QRL.27] | |
| Evaluated traffic index | UK | Annual | [QRL.27] | Ten year series in the Supplementary Statements. 'Evaluated traffic' is a consolidated measure of mails volume reflecting the different workloads in handling letters and parcels - see 3.5.1 |
| Turnover and profit (loss) of overseas mails services | UK | Annual | [QRL.27] | |
| **Postal orders** Number and value issued | UK | Annual | [QRL.27] | 3.4.1 |
| Turnover and profit (loss) | UK | Annual | [QRL.27] | |
| *Counter services* Services By gross sales of postage stamps; postal orders issued; value of transactions on behalf of: National Girobank, National Savings, D.H.S.S., other services | UK | Annual | [QRL.27] | 1.2.1 Ten year operational summary table in the Supplementary Statements |
| **Postal services** By gross sales of postage stamps, local services, retail items sold and postal orders issued. | UK | Annual | [QRL.27] | See the annual operational statistics (counter services) table |
| **National Girobank services** By transcash, withdrawals, local authority rents, DHSS orders, other Girobank Services. | UK | Annual | [QRL.27] | See the annual operational statistics (counter services) table |
| **National Savings services** By NHS deposits and withdrawals, national savings certificates issued and repaid, premium bonds issued and repaid, other savings services | UK | Annual | [QRL.27] | See the annual operational statistics (counter services) table |

| Type of data | Breakdown | Area | Frequency | QRL Publication | Text Reference |
|---|---|---|---|---|---|
| Pensions and Allowances | By child benefits paid, national insurance pensions etc paid, Services allowances, other pensions and allowances. | UK | Annual | [QRL.27] | See the annual operational statistics (counter services) table |
| BT services | By telephone accounts paid, telephone savings stamps sold, telegrams | UK | Annual | [QRL.27] | See the annual operational statistics (counter services) table |
| Other Agency services | By TV licences issued, TV savings stamps sold, CB radio licences issued, motor vehicle licences issued, motor vehicle savings stamps sold, British visitors passports issued, Inland Revenue stamps sold, local tax licences issued, water savings stamps sold, home help savings stamps sold, National Insurance stamps sold. | UK | Annual | [QRL.27] | See the annual operational statistics (counter services) table |
| Percentage of total sub post office business | By type of transaction | UK | 1980 | [QRL.23] | |
| *Telecommunication Services* | | | | | |
| *Telephone calls* | | | | | |
| Number of originated effective inland calls | Local; trunk,by manually controlled and dialled. | UK | Annual | [QRL.31] | 3.4.2 A call is effective when it is answered at the other end. A local call is one between customers who are situated in the same or adjacent charging groups. |

| | | | | | |
|---|---|---|---|---|---|
| Number of originated effective international calls | Maritime, all by outward, inward and transit. | UK | Annual | [QRL.31] | A trunk call is one between customers situated in charging groups that are not adjacent. |
| Percentage of international calls dialled | | UK | Annual | [QRL.31] | |
| Annual calling rate (originated effective calls per station) | By local, inland trunk and international | UK | Annual | [QRL.31] | A station is a telephone set provided for the use of a customer or renter |
| Annual calling rate (originated effective calls) per connexion | By local, inland and international | UK | Annual | [QRL.31] | Each exclusive exchange line and each shared line counts as one exchange connexion |
| Number of calls to guidelines | By Discline; FT Cityline; Traveline; Recipeline; Bedtime Stories; Timeline; Weatherline; Leisureline; Woolworth Gardening; Cricketline; Raceline; Sportsline; localine. | UK | Annual | [QRL.31] | |
| Trunk calls | Distribution of inland manually controlled Trunk Calls over Charge Distance Steps: up to 56Km; over 56Km. | UK | Annual | [QRL.31] | |
| | Distribution of inland customer Dialled Trunk Calls over Charge Distance Steps: up to 56Km; over 56Km | UK | Annual | [QRL.31] | |
| Financial data | Turnover, profit (loss) and return on capital employed for the inland telephone service, by type of rental (business, residential and apparatus), subscribers calls, call office receipts, and Private Circuits | UK | Annual | [QRL.3] [QRL.31] | 3.8.2 and 3.9.2 |

| Type of data | Breakdown | Area | Frequency | QRL Publication | Text Reference |
|---|---|---|---|---|---|
|  | Turnover, profit (loss) and return on capital employed for the international telephone service | UK | Annual | [QRL.3] [QRL.31] | 3.8.2 and 3.9.2 |
| Telegraphs | Number of inland telegrams by source: Public Service by method of acceptance (public counter, phonograms, printergrams) and by type of telegram (ordinary, greetings, overnight), Press, Railways, BT services | UK | Annual | [QRL.31] | 3.4.2 until 1982 |
|  | Number of international telegrams by carrier with international administration and cable companies (onward, inward, transit), Maritime (outward), Radio telegrams (inward), telegrams with Irish Republic (outward, inward) | UK | Annual | [QRL.31] |  |
| Telex | Inland effective traffic: by number of manually controlled calls, dialled calls and dialled metered units | UK | Annual | [QRL.31] | 3.4.2 |
|  | International chargeable traffic: by number of manually controlled calls, Dialled calls (ticketed, metered); number of minutes (outward, inward, transit). | UK | Annual | [QRL.31] |  |

**Productivity** (See also Quality of Service Statistics)

*Postal Services*

| Type of data | Breakdown | Area | Frequency | QRL Publication | Text Reference |
|---|---|---|---|---|---|
| Mails man-hour productivity | Evaluated traffic divided by mails operating traffic | UK | Annual | [QRL.27] | 3.5.1 Six year series in performance indicators 1967-1979, see |
|  | Inward letter work; outward letter work; delivery work. | Inner & Outer London, and other regions | 1980; annually for Inner London only. | [QRL33] [QRL.27] | 3.5.1. Since 1980 there has been an |

| Description | Coverage | Frequency | Reference | Notes |
|---|---|---|---|---|
| | | | | annual index for Inner London mails productivity, published in [QRL.27]. |
| Indices of mail volume handled per worker, compared with a mail volume index. Implicit index of total factor productivity. | UK | Irregular | [QRL.35] | 1922-39 (1922=100), and 1965-76 (1965=100). See 3.5.1 and 2.4.2 |
| Annual percentage change in the cost of producing one unit of output (adjusted for inflation). | UK | Annual | [QRL.27] | See Real Unit Costs table in 'Performance Indicators'. |
| Index of administrative labour cost per mail unit, as compared with an index of mails operations labour costs per mail unit. | UK | Irregular | [QRL.35] | 1965-1976 (1965=100) |
| *Telecommunication Services* Implicit index of total factor productivity - ie changes in volume of business | UK | Annual | [QRL.3] | 3.5.2 Ten year index presented |
| **Quality of Service** *Postal Services* Inland letter service | Percentage of 1st class letters delivered by the working day following collection: by letters posted in the UK and delivered to specified regions, by letters delivered in the UK and posted in specified regions, and by letters posted and delivered in specified regions. | UK and regions | Quarterly | [QRL33] [QRL.27] [QRL33] [QRL.30] | 3.6 Normally the figures cover the last quarter, the last twelve months, and for comparison, the same quarter of the previous year. [QRL.27] gives national averages only, ie no regional breakdown |

| Type of data | Breakdown | Area | Frequency | QRL Publication | Text Reference |
|---|---|---|---|---|---|
| | Percentage of 2nd class letters delivered by the third working day following collection (breakdown as for 1st class letters). | UK and regions | Quarterly | [QRL.33] [QRL.27] [QRL.33] [QRL.30] | [QRL.30] gives performance trends highlighting the London Postal Region. |
| | Christmas Quality of Service, 1st and 2nd class, by percentage of letters delivered on 1st, 2nd, 3rd, 4th and subsequent days. | UK | Dec.1979 | [QRL.30] | See Annex A to [QRL.30] |
| Overseas letter service | Cumulative percentage of air letter mail received from France, in France, on days 2nd, 3rd, 4th and 5th, respectively; by Office of Exchange and 'as delivered'. | UK and France | 1980 | [QRL.31] | |
| | Cumulative percentage of business letter mail delivered abroad by days 2nd/3rd, 4th, 5th, 6th and 7th respectively, for a variety of countries | UK and overseas | 1981 | [QRL.31] | |
| Letter packet, rebate mail and parcels | Regular statistics not available. *Ad hoc* sources only | | | | |
| Letter packets | Percentge 1st class delivered on next working day, and percentage 2nd class delivered on third working day, by 'machine stamped', 'hand stamped' letter packets. | UK | 1980 data only | [QRL.30] | Post Office data via Letter Information service, and also data from a POUNC sample of 4 major letter packet users. |
| Rebate mail | Percentage within the Post Offices 7 day delivery target. | UK | 1980 data only | [QRL.30] | POUNC sample of 4 major users, Sept 1979-April 1980. |

| | | | | | |
|---|---|---|---|---|---|
| Parcels | Average number of days taken to deliver, and percentage of sample delivered within the post offices 2 day delivery target. | UK | 1980 data only | [QRL.30] | POUNC sample of 35,000 items |
| | Percentage of air parcels delivered within 7 days/10 days; percentage of surface parcels delivered within 2 weeks/3 weeks. | UK and West Germany | 1981 | [QRL.31] | |

*Telecommunication Services*
*Telephones*

| | | | | | |
|---|---|---|---|---|---|
| Percentage of calls connected/failed | By cause of failure (percentage that fail due to BT, percentage not connected satisfactorily, percentage 'engaged' or 'no reply') and by type of service (local automatic, STD automatic). | UK and regions | Annual | [QRL.32] [QRL.3] [QRL.37] | 3.6.2 The most comprehensive data is available in [QRL.32], though at the time of writing this 'green' book is under threat of discontinuation. |
| Percentage of inland telephone operator service calls answered within 15 seconds. | | UK | Annual | [QRL.32] [QRL.3] [QRL.37] | |
| Percentage of international telephone operator service calls answered within 15 seconds. | | UK | Annual | [QRL.32] [QRL.3] [QRL.37] | |
| Percentage of international automatic telephone service calls connected/failed | By cause of failure | UK | Annual | [QRL.32] [QRL.3] [QRL.37] | |
| Yearly fault reports per telephone | | UK and regions | Annual | [QRL.32] [QRL.3] [QRL.37] | |
| Percentage of faults reported cleared by end of next working day | | UK and regions | Annual | [QRL.32] [QRL.3] [QRL.37] | |

| Type of data | Breakdown | Area | Frequency | QRL Publication | Text Reference |
|---|---|---|---|---|---|
| Waiting list and supply of service, numbers waiting 2 months or longer for telephone service | | UK | Annual | [QRL.3] [QRL.37] | |

**User Characteristics**

*Postal Services*

| Type of data | Breakdown | Area | Frequency | QRL Publication | Text Reference |
|---|---|---|---|---|---|
| Annual data on average weekly household expenditure on postage | By households with differing ranges of income, by composition and income of household, by type of Administrative Area, and by region. Expenditure on 'postage' can also be broken down by occupation and age of the head of household, and by tenure of household. | UK, regions and Administrative Areas | Annual | [QRL.9] | 2.3.4 Available for all years since 1961 on request to the Department of Employment |
| Volume of letters | By source of origin and destination (e.g. business-business, business-residential, etc.) | UK | Monthly | [QRL.18] | 2.1.3 |
| | By nature of contents, (eg greetings cards, other social, invoice, payment, etc.) | UK | Monthly | [QRL.18] | 2.1.3 |

*Telecommunication Services*

| Type of data | Breakdown | Area | Frequency | QRL Publication | Text Reference |
|---|---|---|---|---|---|
| Annual data on average weekly household expenditure on 'telephone account', and 'telephone, other than account, telegrams, cables'. | Broken down by households with differing ranges of income, by occupation and age of head of household, by tenure of household, by composition and income of household, by type of administrative area and by region. | UK, regions and Administrative Areas | Annual | [QRL.9] | 2.3.4 Available for all years since 1961 on request to the Department of Employment |
| Numerical distribution of telephones | By household composition and by tenure of dwelling | UK, regions and Administrative Areas | Annual | [QRL.9] | 2.3.4 Available for all years since 1961 on request to the Department of Employment |
| Percentage of households with telephone | By socio-economic group of the head of household. | UK | Annual | [QRL.11] | 2.3.6 |

| | | | | | |
|---|---|---|---|---|---|
| Percentage of households with telephones | By possession of a variety of consumer durables, and by readership of a wide range of publications. | UK | Annual | [QRL.20] | |
| Percentage of households with telephone. | By region | UK and regions | Annual | [QRL.34] | 2.3.11 |
| Number of telephone exchange connexions | By type (business, residence, call office, rented coinbox, BT service & connexions). | UK | Annual | [QRL.16] [QRL.42] | See above under 'Output' heading. See above under 'Output' heading. |
| Number of inland telegrams | By type of telegram and source | UK | Annual | [QRL.42] | See above under 'Output' heading. |
| Number of inland and international Telex calls | By type | UK | Annual | [QRL.42] | See above under 'Output' heading. |
| Price elasticity of demand coefficients for telephones | A review of estimates from UK inland and international demand studies. | UK and overseas | 1981 | [QRL.15] | 2.4.3 |

**Final Demand Purchases**

*Postal Services*

| | | | | | |
|---|---|---|---|---|---|
| Consumers expenditure on postal services at current and constant prices. | | UK | Annual | [QRL.19] | 2.3.8 Ten year series |
| Value of exports | By 'surface mail, parcel post and transit', and 'air mails', separately. | UK | Quarterly | [QRL.44] | 2.3.14 Since 1973 on request to the C.S.O. Half yearly data back to 1968 |

*Telecommunication Services*

| | | | | | |
|---|---|---|---|---|---|
| Consumers expenditure on 'telephone and telegraph services' | | UK | Annual | [QRL.19] | 2.3.8 Ten year series |
| Value of exports | By 'overseas telegraph', 'telex', 'telephone', 'international leased circuits', 'submarine cables and satellite projects', separately. | UK | Quarterly | [QRL.44] | 2.3.14 Since 1973 on request to the C.S.O. Half year data back to 1968 |

*Postal and Telecommunication Services*

| | | | | | |
|---|---|---|---|---|---|
| Current expenditure by general Government on goods and services of MLH 708 | | UK | Annual | [QRL.19] | 3.8.1 Available on request to the C.S.O. |

| Type of data | Breakdown | Area | Frequency | QRL Publication | Text Reference |
|---|---|---|---|---|---|
| Gross Domestic Capital Formation | By type of asset | UK | Annual | [QRL.19] | 3.8.1 |
| **Intermediate Demand Purchases** | | | | | |
| *Postal and Telecommunication Services* | | | | | |
| Annual expenditure by each MLH industry on the item 'postage, telephone, telegrams, cables and telex', for establishments employing over 300 persons. | | UK | 1948, 1963 and 1968 only | [QRL.5] | 2.3.3 and 3.8.1 |
| Intermediate purchases of 'postal and telecommunications services'. | By industry group. | | Full information tables 1954, 1963, 1968 and 1974 | [QRL.14] | 2.3.7 102 industry/commodity groups separately identified in the 1974 input-output tables |
| Direct and indirect import of industrial output, in coefficient form. | Total and direct requirements of imported commodities per 1000 units of domestic output. | | Full information tables 1954, 1963, 1968 and 1974 | [QRL.14] | 2.3.7 102 industry/commodity groups separately identified in the 1974 input-output tables |
| **Labour Statistics** | | | | | |
| *Postal Services* | | | | | |
| Number and type of persons employed | Total employed: by business location and function (Postal Headquarters, Regional Headquarters, National Television Licence Records Office, Philatelic Bureau, Post Offices, Mails Operations, Engineering, Motor Transport, Supplies Department); full and part time; Subpostmasters employed on an agency basis. | UK | Annual | [QRL.27] | Ten year time series presented in the Supplementary Statistics |

| | | | | |
|---|---|---|---|---|
| Postman and postman higher-grade (mails services): gross hours, overtime hours. Postal officer and postal assistant (counter services): gross hours, overtime hours. | UK | Annual | [QRL.27] | Operational Statistics Table |
| Total employed: by employment status and by sex; by occupation and by sex. | GB; Scotland | 1981, and Decennial | [QRL.4] | 2.3.2 |
| *Telecommunication Services* Total employed in BT | | | | |
| By business location (Telecom Headquarters, Regional Headquarters, Telephone Areas), and function (General Management, Personnel, Finance & Management Services, Planning & Works, Services & Marketing, R & D, Procurement, Accommodation, Motor Transport, Data Processing). | UK | Annual | [QRL.42] [QRL.3] | |
| By employment status and by sex; by occupation and by sex. | GB; Scotland | 1981 | [QRL.4] | 2.3.2 |
| *Postal and Telecommunication Services* Total employed | | | | |
| By region; by sex; by full and part time. | UK | Monthly | [QRL.8] | From 1983, as this source changes to the 1980 SIC (see 1.4.1), Posts and Telecomm-unications should be available separately |
| By industrial status and by sex; by occupational type and by sex; by socio-economic class and sex; by age and marital status; by sex and area of workplace. | UK | 1981, and Decennial | [QRL.4] | |
| **Wage Rates and Earnings** | | | | |
| *Postal Services* Total employee compensation | | | | |
| Pay, pensions, employers social security contributions, subpostmasters remunerations. | UK | Annual | [QRL.27] | |

| Type of data | Breakdown | Area | Frequency | QRL Publication | Text Reference |
|---|---|---|---|---|---|
| Wage rates | By occupational type (postmen - by grade, postal officers, labourers, technicians, technical officers); supplements, guaranteed payments, rates for overtime; for young workers (less than 20 years), and paid holiday entitlement | UK | Annual | [QRL.43] | |
| | 'Postmen, mail sorters and messengers': gross weekly average earnings, gross hourly earnings and pay at various percentiles of the income distribution for this occupational category. | UK | Annual | [QRL.21] | |
| *Telecommunication Services* | | | | | |
| Total employee compensation in BT | By number of employees in various salary ranges. | UK | Annual | [QRL.3] | |
| Wage rates | By occupational type (telegraphists, telephonists, etc.) | UK | Annual | [QRL.43] | |
| *Postal & Telecommunication Services* | | | | | |
| Earnings | Median, quartiles & deciles of gross weekly/hourly earnings, by sex and MLH industry; percentage with earnings 'less than' specified amounts. | UK | Annual | [QRL.21] | |
| **Fixed Assets** | | | | | |
| *Postal Services* | | | | | |
| Additional assets | Net additions to plant and machinery, vehicles, computers, other equipment, as well as land and buildings. Ad hoc data on the Mechanised Letter Office programme, eg number of fully mechanised letter and parcels offices. | UK | Annual | [QRL.27] | See also [QRL.24] for various discounted cash flow costings of the mechanisation plan |

| | | | | | |
|---|---|---|---|---|---|
| Gross replacement cost, accumulated depreciation, and net replacement cost for posts | By type of asset (land and buildings, by freehold, long leasehold, short leasehold; plant and machinery; vehicles, fixtures and fittings; and assets in course of construction). | UK | Annual | [QRL.27] | Notes on the Financial Statements. Also ten year series for value of tangible fixed assets in Supplementary Statements. |
| Gross historical cost | By type of asset, and Net Book Value, by type of asset. | UK | Annual | [QRL.27] | Notes on the Financial Statements. Also ten year series for value of tangible fixed assets in Supplementary Statements. |
| Number of delivery points; number of Post Offices at year end (Crown offices, Scale Payment Sub-offices). | | UK | Annual | [QRL.27] | Notes on the Financial Statements. Also ten year series for value of tangible fixed assets in Supplementary Statements. |
| *Telecommunication Services* | | | | | |
| Gross replacement cost, accumulated depreciation and net replacement cost for BT | By type of asset (land and buildings; plant and equipment) | UK | Annual | [QRL.3] | |
| Net fixed assets at historical cost, and at replacement cost. | | UK | Annual | [QRL.3] | Ten year series |
| *Telephone Exchanges* | | | | | |
| Telephone exchanges | Classified into types (Strowger, Crossbar, Electronic - director, non-director, etc.) and sizes (ie number of connexions). | UK | Annual | [QRL.42] | |

| Type of data | Breakdown | Area | Frequency | QRL Publication | Text Reference |
|---|---|---|---|---|---|
| Total net capacity | By type of exchange; spare capacity by type of exchange. | UK | Annual | [QRL.42] | |
| Types and sizes of telephone exchanges | By telephone area, and locality (by charge group). | UK, telephone areas and localities | Annual | [QRL.16] | |
| *Exchange connexions* Total number, new, ceased and net annual increase | | UK | Annual | [QRL.42] | |
| Number of exchange connexions | By type: business, residence, call office, BT service, rented coinbox. | UK | Annual | [QRL.42] | |
| Number of exchange connexions | By type (business, residence), by telephone area and locality (by charge group). | Telephone area and localities | Annual | [QRL.16] | |
| Number of shared service connexions | By type: business, residence | UK | Annual | [QRL.42] | |
| Public telephone call offices | By type (kiosks, cabinets, open call offices) and by location (urban, rural). | UK | Annual | [QRL.42] | |
| Number of exchange connexions served by each type of exchange | | UK | Annual | [QRL.42] | |
| *Telephone Stations* Total number, new, ceased and net annual increase; ratio of stations to connexions at year end. | | UK | Annual | [QRL.42] | |
| Number of stations | By type: business, residence, call office, BT and service, private circuit, radiophones | UK | Annual | [QRL.42] | |
| Number of stations served | By type of exchange (manual, automanual, automatic - by type). | UK | Annual | [QRL.42] | |
| Number of stations | By telephone area, locality and charge group. | Telephone area and localities | Annual | [QRL.16] | |

| | | | | | |
|---|---|---|---|---|---|
| Number of stations | By type; by telephone area | Telephone areas | Annual | [QRL.16] | |
| *Equipment and Line Plant* | | | | | |
| Additions to, and net expenditure on, inland equipment | Transmission equipment (duct, cable, radio and repeater equipment); exchange equipment (Strowger, Crossbar, semi-electronic, digital); customer equipment; office machines and furniture; motor vehicles; and miscellaneous equipment. | UK | Annual | [QRL.3] | |
| Additions to, and net expenditure on, international cables and equipment. | | UK | Annual | [QRL.3] | |
| Net book values of inland service plant | By type (trunk and junction circuits; local lines; customers circuits; exchange equipment; telegraph equipment; Datel equipment). Net book value of international service plant. Net book value of accommodation by type (land, buildings, heating and lighting plant, furniture), and of motor transport, office machines. | UK | Annual | [QRL.42] | Details ceased in 1981 |
| Customers cables | Number by type of connexion | UK | Annual | [QRL.42] | |
| Telephone trunk lines in use | Number by type of service (public, private) | UK | Annual | [QRL.42] | |
| Motor transport | Number of vehicles by type, and fleet mileage. | UK | Annual | [QRL.42] | |
| *Telegraph Equipment* | | | | | |
| Telegraph multi-channel voice-frequency systems, and circuits in use | By type of circuit. | UK | Annual | [QRL.42] | |
| *Telex* | | | | | |
| Number of exchanges, trunk lines and exchange connexions | By type | UK | Annual | [QRL.42] | |

| Type of data | Breakdown | Area | Frequency | QRL Publication | Text Reference |
|---|---|---|---|---|---|
| *Datel* | | | | | |
| Number of modems | | UK | Annual | [QRL.42] | |
| **Current Assets** | | | | | |
| *Postal Services* | | | | | |
| Values, and changes, in stocks, debtors, investments and cash at bank and in hand. | | UK | Annual | [QRL.27] | 3.12.1 See Current Cost Balance Sheet and Current Cost Source and Application of Funds Table. Ten year series for value of current assets in Supplementary Statements |
| *Telecommunication Services* | | | | | |
| Values, and changes in, stocks, debtors and cash at bank and in hand. | | UK | Annual | [QRL.3] | 3.12.2 Balance Sheet and Source and Application of Funds Table |
| **Liabilities** | | | | | |
| *Postal Services* | | | | | |
| Value of capital and reserves | Long term loans from the Secretary of State, by interest rate and length of loan; foreign loans, by interest rate and length of loan; revaluation reserve; profit and loss account; other reserves. | UK | Annual | [QRL.27] | See Current Cost Balance Sheet and Notes on the Financial Statements |
| Amount owed to creditors falling due within one year | Agency service balance; inter business balances; loans; other creditors. Amount owed to creditors falling due beyond one year. | UK | Annual | [QRL.27] | See Current Cost Balance Sheet. Ten year series in Supplementary Statements. |
| *Telecommunication Services* | | | | | |

| | | | | | |
|---|---|---|---|---|---|
| Creditors, amounts falling due within one year | By type of creditor | UK | Annual | [QRL.3] | |
| Loan capital and reserves | By type | UK | Annual | [QRL.3] | See Balance Sheet and Notes to Supplementary Current Cost Statements |
| Long term liability to Post Office Staff Superannuation Scheme | | UK | Annual | [QRL.3] | |

**Other Financial Statements**

*Income, expenditure and profit*

*Postal Services*

| | | | | | |
|---|---|---|---|---|---|
| Turnover | By type of service | UK | Annual | [QRL.27] | |
| Profit | By type of service | UK | Annual | [QRL.27] | |
| Costs | By type: staff costs (wages and salaries; social security costs; pension costs; pension fund deficiency contribution); other operating charges (subpostmasters remuneration; auditors remuneration; leasing; emoluments of members; retirement pensions and gratuities); depreciation and other amounts written off tangible fixed assets. | UK | Annual | [QRL.27] | See Current Cost Profit and Loss Account, and Notes on the Financial Statements for further detail |
| Cost allocations for letter mail services | By functions | UK | 1982 | [QRL.24] | |
| Achievement against profit target (percentage on turnover) | | UK | Annual | [QRL.27] | See Performance Indicators |

*Telecommunication Services*

| | | | | | |
|---|---|---|---|---|---|
| Turnover | By type of service | UK | Annual | [QRL.3] [QRL.42] | |
| Additional turnover due to price changes; to business expansion | | UK | Annual | [QRL.3] | Ten year series presented |
| Profit | By type of service | UK | Annual | [QRL.3] | |

| Type of data | Breakdown | Area | Frequency | QRL Publication | Text Reference |
|---|---|---|---|---|---|
| Operating costs | By type | UK | Annual | [QRL.3] [QRL.42] | See table in [QRL.14] on details of operating costs |
| Additional operational costs due to changes in pay and price levels | | UK | Annual | [QRL.3] | Ten year series |
| Capital expenditure | By type | UK | Annual | [QRL.3] | See table on details of capital expenditure |
| Growth in capital charges, net of pay and price change | | UK | Annual | [QRL.3] | Ten year series |
| Growth in other costs, net of pay and prices | | UK | Annual | [QRL.3] | Ten year series |
| Achievement against profit target (percentage of capital employed at replacement cost). | | UK | Annual | [QRL.3] | Ten year series |
| *Sources of Finance* | | | | | |
| *Postal Services* | | | | | |
| Funds from internal sources | By type: profit; net interest payable (receivable); depreciation and other amounts written off tangible fixed assets: monetary working capital adjustments. | UK | Annual | [QRL.27] [QRL.10] | See Current Cost Source and Application of Funds table, and Notes on the Financial Statements in [QRL.27]. |
| Funds from external sources | By type: loans from the Secretary of State; Public Dividend Capital; Foreign loans. | UK | Annual | [QRL.27] [QRL.10] | See Current Cost Source and Application of Funds table, |

| | | | | | |
|---|---|---|---|---|---|
| Achievement against target for the External Financing Limit settled with the Government. | | UK | Annual | [QRL.27] | and Notes on the Financial Statements in [QRL.27] See Performance Indicators for a six year table |
| *Telecommunication Services* | | | | | |
| Funds from internal sources | Retained profit; depreciation | UK | Annual | [QRL.3] [QRL.27] [QRL.42] [QRL.10] | Ten year series also presented in [QRL.3] |
| Funds from external sources | By type | UK | Annual | [QRL.3] [QRL.27] [QRL.42] [QRL.10] | Ten year series also presented in [QRL.3] |
| **Prices** *Postal Services* Tariff index for 'postage' in General Retail Price Index | | UK | Monthly | [QRL.8] | |
| Tariff index adjusted for inflation, mails services | | UK | Annual | [QRL.27] | Ten year series in Supplementary Statements 3.7.1 |
| Report on each package of proposed changes in tariffs | | UK | Irregular | [QRL.38] | |
| Weighting of postal services in General Retail Price Index and Pensioner Households Retail Price Index | | UK | Annual | [QRL.8] | |
| *Telecommunication Services* Tariff index for 'telephones and telemessages' in General Retail Price Index | | UK | Monthly | [QRL.8] | |

| Type of data | Breakdown | Area | Frequency | QRL Publication | Text Reference |
|---|---|---|---|---|---|
| Tariff indices, adjusted and unadjusted for inflation | | UK | Annual | [QRL.3] | Ten year series in Supplementary Statements |
| Principal changes in general inland telephone and telegraph tariffs since 1912 | | UK | Annual | [QRL.42] | See Historical Section |
| Principal changes in general inland telex tariffs since 1954 | | UK | Annual | [QRL.42] | See Historical Section |
| Comparison of Retail Price Index with Telecommunications Tariff Index | | UK | Annual | [QRL.37] | Ten year series; Also comparison made between Telecommunications Tariff Index and other selected items over past ten years See 2.3.12. Ten year series presented |
| Ratio of the relative price of telephone calls and mail | | UK | Annual | [QRL.39] | |
| Weighting of Telecommunications Services in General Retail Price Index and Pensioner Households Retail Price Index | | UK | Annual | [QRL.8] | |

**Some International Comparisons**

*Postal Services*

| Type of data | Breakdown | Area | Frequency | QRL Publication | Text Reference |
|---|---|---|---|---|---|
| Number of post offices per 20,000 of population. comparison | An international comparison | UK and overseas | 1983 | [QRL.27] | |

| | | | | | |
|---|---|---|---|---|---|
| Cost of an inland letter | Using tariffs adjusted to reflect the purchasing power of currencies. | UK and overseas | 1983 | [QRL.27] | |
| International comparison of postal rates; also of delivery and Sunday Collection Services. | | UK and overseas | March 1980 | [QRL.23] | |
| Comparative price data | By service type | UK and overseas | Irregular | [QRL.38] | See 3.7.1 |
| *Telecommunication Services* | | | | | |
| Typical UK single line residential bills compared with overseas tariffs | | UK and overseas | Annual | [QRL.37] | |
| Typical UK single line business bills compared with overseas tariffs | | UK and overseas | Annual | [QRL.37] | |
| International comparison of tariffs for 100 km telephone circuits | | UK and overseas | 1981 | [QRL.15] | |
| Measures of data transmission penetration | By country, together with forecast | UK and overseas | 1981 | [QRL.15] | |
| Review of price elasticity coefficients | | UK and overseas | 1981 | [QRL.15] | |

## QUICK REFERENCE LIST KEY TO PUBLICATIONS

| Reference | Author or Organisation | Title | Publisher | Frequency or date | Remarks |
|---|---|---|---|---|---|
| [QRL.1] | Department of Employment | British Labour Statistics Yearbook | H.M.S.O. | Annual | |
| [QRL.2] | British Telecom | British Telecom Guide | H.M.S.O. | Annual | |
| [QRL.3] | British Telecom | British Telecom Report and Accounts | British Telecom | Annual (usually July) | |
| [QRL.4] | Office of Population Censuses and Surveys | Census of Population | H.M.S.O. | Decennial | |
| [QRL.5] | Department of Industry Business Statistics Office | Census of Production | H.M.S.O. | Annual | See (2.3.3) |
| [QRL.6] | Department of Employment | Changes in Rates of Wages and Hours of Work | H.M.S.O. | Monthly | |
| [QRL.7] | City of Kingston upon Hull Telephone Department | City of Kingston upon Hull Telephone Department Statistics | See contact point in Appendix 1 | Irregular | |
| [QRL.8] | Department of Employment | Employment Gazette | H.M.S.O. | Monthly | |
| [QRL.9] | Department of Employment | Family Expenditure Survey | H.M.S.O. | Annual | |
| [QRL.10] | C.S.O. | Financial Statement and Budget Report | H.M.S.O. | Annual | |
| [QRL.11] | Office of Population Censuses and Surveys | General Household Survey | H.M.S.O. | Annual | |
| [QRL.12] | Post Office Users National Council | How Good is our Postal Service? | POUNC | March 1972 | See [QRL.29] for address |
| [QRL.13] | The Monopolies and Mergers Commission | The Inner London Letter Post | H.M.S.O. | March 1980 | |
| [QRL.14] | Department of Industry, Business Statistics Office | Input-Output Tables for the UK | H.M.S.O. | Irregular see (2.3.7) | |
| [QRL.15] | Department of Industry | Liberalisation of the use of British Telecommunications Network, A study by M.E. Beesley | H.M.S.O. | Jan 1981 | |
| [QRL.16] | British Telecom | List of Exchanges | British Telecom | Annual | |
| [QRL.17] | Post Office | London Post Offices and Streets | H.M.S.O. | Annual | |

| Ref | Organization | Title | Publisher | Date/Frequency | Notes |
|---|---|---|---|---|---|
| [QRL.18] | The Post office | *The Mail Classification Survey*; now integrated within the Letter Information System | Unpublished data, for which application may be made to the Post Office contact point in Appendix 1 | Monthly | See (2.1.3) below |
| [QRL.19] | C.S.O. | *National Income and Expenditure Blue Book* | H.M.S.O. | Annual | |
| [QRL.20] | Joint Industry Committee for National Readership Surveys (JICNARS) | *National Readership Survey* | JICNARS | Annual | |
| [QRL.21] | Department of Employment | *New Earnings Survey* | H.M.S.O. | Annual | |
| [QRL.22] | The Select Committee on the Nationalized Industries | *The Post Office* | H.M.S.O. | Feb 1967 | |
| [QRL.23] | House of Commons Industry and Trade Committee | *The Post Office: First Report from the Industry and Trade Committee* | H.M.S.O. | March 1980 | |
| [QRL.24] | House of Commons Industry and Trade Committee | *The Post Office: Fifth report from the Industry and Trade Committee* | H.M.S.O. | April 1982 | |
| [QRL.25] | Post Office | *Post Office Guide* | H.M.S.O. | | |
| [QRL.26] | The Select Committee on the Nationalized Industries | *The Post Office's Letter Services* | H.M.S.O. | 1975 | |
| [QRL.27] | Post Office | *Post Office Report and Accounts* | The Post Office | Annual (usually July) | |
| [QRL.28] | Post Office | *Post Office in the UK* | H.M.S.O. | | |
| [QRL.29] | Post Office | *Postal Addresses & Index Postcode Directories* | H.M.S.O. | | |
| [QRL.30] | Post Office Users National Council. | *The Postal Service: A Further Study Report No.23* | POUNC. | Oct.1980 | |
| [QRL.31] | Post Office Users National Council. | *The Postal Service: Overseas Mail Report No.28* | POUNC. | Dec 1982 | |
| [QRL.32] | British Telecom | *Quality of Telephone Services* | British Telecom | Annual (usually July) | Quarterly updates available on request |
| [QRL.33] | The Post Office | *Quality of the Inland Letter Service* | The Post Office | Quarterly | |
| [QRL.34] | The Post Office | *Regional Trends* | H.M.S.O. | Annual | |

| Reference | Author or Organisation | Title | Publisher | Frequency or date | Remarks |
|---|---|---|---|---|---|
| [QRL.35] | Department of Industry | Report of the Post Office Review Committee Chairman C.F. Carter | H.M.S.O. Cmnd 6850 | 1977 | |
| [QRL.36] | Post Office Users National Council | Report on the delivery, performance and potential of the Post Office's Mail Services, Report No.17 | POUNC. Waterloo Bridge House Waterloo Road London SE1 8UA | Jan 1979 | |
| [QRL.37] | British Telecom | A Report to Customers | British Telecom | Annual | Based on [QRL.3] |
| [QRL.38] | Post Officers Users National Council | Reports on each package of proposed price changes | POUNC. Waterloo Bridge House Waterloo Road London SE1 8UA | Irregular | |
| [QRL.39] [QRL.40] | C.S.O. States of Jersey Telecommunications Board | Social Trends States of Jersey Telecommunications Board - Annual Report | H.M.S.O. See Contact point designated in Appendix 1 | Annual Annual | |
| [QRL.41] | Department of Employment | Strikes in Britain, Manpower Paper No.15 | Department of Employment | 1978 | |
| [QRL.42] | British Telecom | Telecommunications Statistics | British Telecom | Annual | The Orange Book |
| [QRL.43] | Department of Employment | Time Rates of Wages and Hours of Work | H.M.S.O. | Annual | |
| [QRL.44] [QRL.45] | C.S.O. Consumers Association | UK Balance of Payments Which? | H.M.S.O. Consumers Association | Annual Monthly | |

# BIBLIOGRAPHY

[B.1] M.Corby, *The Postal Business*, Kogan Page, 1979

[B.2] A.Dean, "Wages and Earnings" in *Reviews of UK Statistical Sources* Vol.13

[B.3] A.Hazelwood, "Telecommunication Statistics", in *The Sources and Nature of the Statistics of the UK*, (Ed. M.G.Kendall) The Royal Statistical Society, 1957

[B.4] W.F.Kemsley, R.U.Redpath, and M.Holmes, *Family Expenditure Survey Handbook*, Social Survey Division of OPCS, H.M.S.O., 1980

[B.5] R.Maurice, *National Accounts Statistics, Sources and Methods* H.M.S.O., 1968

[B.6] D.L.Munby and A.H.Watson, "Road Passenger Transport and Road Goods Transport" in *Reviews of UK Statistical Sources* Vol.7, 1978

[B.7] B.Smith, *Estimating the Number of Local Telephone Calls in the UK*, Diploma dissertation, Central London Polytechnic, 1977

[B.8] *Classification of Occupations and Directory of Occupational Titles*, H.M.S.O., 1972

[B.9] "A Computerised Spare-plant Return - SPRET" in *Post Office Electrical Engineers Journal* Vol.69, Part 3, Oct 1976, p180

[B.10] *Social Trends* 14, Central Statistical Office, 1983.

[B.11] *The Standard Industrial Classification, Revised 1980* H.M.S.O. (For further detail see *Indexes to the SIC, Revised 1980*)

[B.12] "Input-Output Tables for the UK" in *Studies in Official Statistics* No.22

[B.13] *A Study of UK Nationalised Industries* Background Paper No.6, NEDO, 1977

# APPENDIX 1: LIST OF CONTACT POINTS

The single contact points designated for the Post Office and British Telecom respectively, will route all enquiries to the appropriate departments.

**The Post Office**

Financial Planning Department (FP2.3)
Room 118
Armour House
St Martin's le grand
LONDON EC1A 1AR

**British Telecom**

Organisation, Performance & Systems Department
OPS1.2
Leith House
47-57 Gresham Street
LONDON EC2V 7JL

Public Relations Officer
City of Kingston upon Hull Telephone Department
Telephone House
Carr Lane
KINGSTON UPON HULL
HU1 3 RE

States of Jersey Telecommunications Board
Telephone House
JERSEY
Channel Islands

# SUBJECT INDEX

# my revision notes

Pearson Edexcel GCSE (9–1)

# DESIGN AND TECHNOLOGY

Ian Fawcett

Andy Knight

Jacqui Howells

David Hills-Taylor

HODDER
EDUCATION
AN HACHETTE UK COMPANY

Orders: please contact Bookpoint Ltd, 130 Park Drive, Milton Park, Abingdon, Oxon OX14 4SE. Telephone: +44 (0)1235 827827. Fax: +44 (0)1235 400401. Email education@bookpoint.co.uk Lines are open from 9 a.m. to 5 p.m., Monday to Saturday, with a 24-hour message answering service. You can also order through our website: www.hoddereducation.co.uk

ISBN: 9781510480506

© Ian Fawcett, Andy Knight, Jacqui Howells and David Hills-Taylor 2020

First published in 2020 by

Hodder Education,

An Hachette UK Company

Carmelite House

50 Victoria Embankment

London EC4Y 0DZ

www.hoddereducation.co.uk

Impression number   10  9  8  7  6  5  4  3  2  1

Year    2024   2023   2022   2021   2020

Cover photo © Phonlamai Photo/Shutterstock.com

Illustrations by Integra Software Services Ltd.

Typeset in India by Integra Software Services Ltd.

Printed in Spain.

A catalogue record for this title is available from the British Library.

# Get the most from this book

Everyone has to decide their own revision strategy, but it is essential to review your work, learn it and test your understanding. These Revision Notes will help you to do that in a planned way, topic by topic. Use this book as the cornerstone of your revision and don't hesitate to write in it: personalise your notes and check your progress by ticking off each section as you revise.

## Tick to track your progress

Use the revision planner on pages 4–6 to plan your revision, topic by topic. Tick each box when you have:

- revised and understood a topic
- tested yourself
- practised exam questions and gone online to check your answers and complete the quick quizzes.

You can also keep track of your revision by ticking off each topic heading in the book. You may find it helpful to add your own notes as you work through each topic.

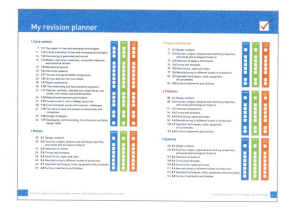

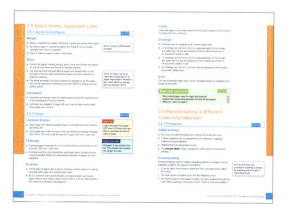

# Features to help you succeed

## Exam tips

Expert tips are given throughout the book to help you polish your exam technique in order to maximise your chances in the exam.

## Typical mistakes

The authors identify the common mistakes candidates make and explain how you can avoid them.

## Now test yourself

These short, knowledge-based questions provide the first step in testing your learning. Answers are given online at **www.hoddereducation.co.uk/ myrevisionnotes**

## Definitions and key words

Clear, concise definitions of essential key terms are provided.

## Exam practice

Practice exam questions are provided at the end of each section. Use them to consolidate your revision and practise your exam skills.

## Online

Go online to check your answers, download the glossary and try out the extra quick quizzes at **www.hoddereducation.co.uk/myrevisionnotes**

# My revision planner

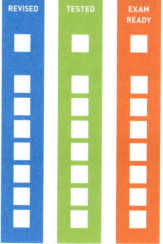

## Answers and glossary at
www.hoddereducation.co.uk/myrevisionnotes

## 1.1 The impact of new and emerging technologies

### 1.1.1 Industry

- Modern factories increasingly use **automated production**. Robots are often used to complete some repetitive and monotonous tasks previously carried out by humans, which can lead to unemployment.
- **Computer-aided manufacture (CAM)** machines are used where high volumes of identical products of a consistent high quality are needed.
- People often move to different countries to find suitable employment or where there is a shortage in a particular workforce. This is referred to as **demographic movement**.
- Science and technology parks are areas where like-minded businesses come together to form centres of excellence.

> **Automated production**: a production method of using machinery controlled by computers.
>
> **CAM**: computer-aided manufacture.

### 1.1.2 Enterprise

- A privately-owned business is owned by the company founders, or their families, or by a small group of investors whose shares in the company are not publicly traded.
- Crowd funding is an internet-based scheme allowing people to raise money to get a business idea started, such as the manufacture of a new product. Investors lend money to projects that they believe are viable.
- The UK government offers some loans and grants to new start-up businesses. Investors can get tax relief on their investments. While not all new businesses are successful, the potential to boost the economy and the provision of future employment is worth taking a risk.
- Not-for-profit organisations reinvest all profits back into the business for the benefit of all involved.

**Figure 1.1 The Forest Stewardship Council® (FSC®) is an example of a not-for-profit organisation**

### 1.1.3 Sustainability

- **Sustainability** is about meeting today's needs without compromising the needs of future generations.
- Demand on the world's natural resources currently exceeds supply. Materials used in product design fall into two categories:
  - **finite**: resources with a limited supply that will eventually run out
  - **non-finite**: resources that can be replenished and are unlikely to run out.
- One way to reduce our environmental impact is to consider the six Rs.

> **Sustainability**: the manufacture of goods without compromising future needs; the materials required for a product can be replenished and will not run out.

**Table 1.1 The six Rs of sustainability**

| | |
|---|---|
| **Rethink** | Is there a better way of making the product that is less harmful to the environment? Can the design be simplified to make manufacturing easier? |
| **Recycle** | Can the product be recycled easily after it is no longer needed? Can the components and materials be separated easily? Can recycled materials be used? |
| **Repair** | Can this product be repaired easily if it breaks? Can the component parts be replaced easily? |
| **Refuse** | Consumers might not buy a product if it is not environmentally friendly or if it is made by people working in poor conditions. |
| **Reduce** | Can the number of component parts, new materials or packaging be reduced? Can the manufacturing process be simplified to reduce the energy used? |
| **Reuse** | Can any parts be reused after it is no longer needed? Could the product be reused for something else once its primary use is no longer required? |

## Transportation

- Each product has a **carbon footprint** from its manufacture and the distance it travels until it is purchased and used.
- Transportation costs/carbon footprint can be reduced by:
  - sourcing materials closer to manufacturing centres and potential markets
  - using energy-efficient means of transporting goods, for example electric or **hybrid vehicles**.

## Pollution

The manufacture and use of any product leads to an increase in pollution levels. There are ways to reduce the level of pollution.

- Manufacturers could use cleaner energy sources, such as solar or wind.
- Goods could be transported using electric vehicles that do not emit greenhouse gases.
- Products could be built to last and easy to repair, causing less waste.
- Products need to be disposed of carefully if they cannot be recycled, avoiding landfill where possible.

## Demand on natural resources

- Natural resources and productive land enable us to produce the goods and services that sustain and support modern lifestyles.
- The **ecological footprint** is a measure of the impact that human activity has on the environment.
- Humanity's ecological footprint is currently 1.7 Earths – the Earth takes about 18 months to replace what we use in 12 months. If this continues, we will create an **ecological deficit**.

> **Carbon footprint**: a measure of the greenhouse gases emitted by human activity.
>
> **Hybrid vehicles**: vehicles that are both petrol and electric, so emissions are reduced.

**Figure 1.2 Hybrid technology reduces carbon emission from vehicles**

> **Ecological footprint**: a measure of the impact that human activity has on the environment.
>
> **Ecological deficit**: when resources are being used up faster than nature can replenish them.

## Waste generated

- Many products are designed with built-in or planned **obsolescence**, meaning that they are not designed to be long lasting.
- Companies such as Apple introduce new products that consumers are eager to own, even though their current products still work perfectly well. This puts further strain on the world's resources, as many products are not recycled or are difficult to recycle.
- Some polymers are difficult to recycle. New technology is being developed that will allow them to be broken down more effectively and safely.

> **Obsolescence**: the process of becoming no longer needed or wanted.

## 1.1.4 People

REVISED

- Developments in mobile technology and the internet make it easier for us to communicate with people all over the world. This leads to greater competition among manufacturers.
- There are downsides to this global society, which affect workers directly:
  - Importing cheap products from overseas rather than buying locally produced products can lead to job losses in our own society.
  - The use of automation leads to job losses (see Section 1.1.1 Industry on page 7).
  - Workers overseas are often paid low wages to reduce costs and maximise profit.
- The workforce of the future needs to be more adaptable, offering different skills, such as problem solving, which lead into more highly skilled jobs like engineering or computer science.
- Technological developments in **computer-aided design (CAD)** packages have changed the way designers work. All aspects of developing design ideas can be done using computers.
- Apprenticeships are work-based training programmes that allow trainees to learn on-the-job from a skilled employer. Apprenticeships are often in vocational areas, for example tailoring, engineering or carpentry.
- Society needs to be inclusive. New and emerging technologies have allowed designers to create products that meet the needs of many people, including those with disabilities and the elderly. This could be from simple hand tools to computer programs.
- Children are increasingly exposed to technology. While this can be positive, many believe it is at the expense of social interaction.

> **CAD**: computer-aided design; software used to produce drawings and virtual models of products and systems.

## 1.1.5 Culture

REVISED

- People have always travelled to find better work opportunities and improve their lives. A permanent move is called **migration**. The European Union allows EU nationals to travel freely across the member states.
- Social segregation can occur when ethnic minorities cluster together to form their own society but do not fully integrate into the wider community.

> **Migration**: the movement of people, often to find work.

## 1.1.6 Society

- Many factories now operate 24/7 with people working in shifts in rotation, including night shifts. Productivity is increased, but at the expense of a work–life balance for many workers.
- People are increasingly reliant on technology. Many devices can be connected to the internet and be controlled remotely, for example controlling the lighting or heating in your home from a smart phone. This is referred to as the **Internet of Things (IoT)**.
- Advances in communication technology allows the use of video conferencing to conduct meetings with people in different locations, enabling remote working and reducing the need to travel.

> **Internet of Things (IoT)**: the interconnection of everyday products to the internet.

## 1.1.7 Environment

- Burning fossil fuels to provide energy releases greenhouse gases into the environment – a major contributor to global warming.
- Developments in renewable energy allow us to make better use of alternative sources of energy, reduces our reliance on **finite fossil fuels** and lowers pollution levels.
- Many companies sell recyclable waste materials to recoup some costs. Recycling schemes allow consumers to dispose of waste safely and thoughtfully, for example collection points for old batteries in many retail outlets.
- Complex products are often difficult to recycle if the materials cannot be separated easily, for example foil-lined cardboard tube packaging from crisps, which have a metal base, metal tear off lid and polymer cap.
- Many consumers believe packaging of goods is unnecessary, particularly if the packaging used is not recycled. Eliminating the need for packaging, reducing the amount used or only using recyclable or **biodegradable** materials is better for the environment.

> **Finite fossil fuels**: fuels that have a limited supply and cannot be replaced, for example, oil and coal.
>
> **Biodegradable**: will decompose (rot) and break down without damaging the environment.

## 1.1.8 Production techniques and systems

Products are manufactured under different scales of production depending on the:

- type of product being made
- quantity required
- timescale for manufacture.

> **Bespoke**: specially made for a particular person.

**Table 1.2 Different scales of production**

| One-off | Batch | Mass | Continuous |
|---|---|---|---|
| A single product is manufactured for a specific client need, for example a **bespoke** suit | A set number of identical products is produced over a set period of time, for example seasonal products | Very high volumes produced over an extended period of time to meet with the demands of mass marketing | High volume of production; manufacturing is non-stop through a 24-hour period |

> **Exam tip**
>
> Reasons for choosing one scale of production over another vary: a uniform style and size of paper clip, for example, would be mass produced as millions are potentially needed and the style is unlikely to change. Know why different scales are used and be able to give reasons to support your answer.

- Many components, parts and materials are made in standardised measurements and sizes, and are readily available for manufacturers to use when needed.
- **Just-in-time (JIT)** manufacturing is a stock control system where components, parts or materials are ordered when needed. This is an efficient, cost-effective system but is highly reliant on suppliers delivering goods on time.
- Lean manufacturing focuses on minimising waste in all areas of design through to manufacture, while maximising productivity at the same time.

> **Typical mistake**
>
> When a question asks you to 'explain' something, the answer requires a fact and a further elaboration of that fact. You will not gain full marks for simply listing features that do not include reasoning.

## Now test yourself

1 Explain how automation used in industry is changing the way products are manufactured. (4)
2 List **two** ways an entrepreneur could gain funding for a new business venture. (2)
3 Give a reason why seasonal products are often batch produced. Include an example of a batch-produced product in your answer. (2)
4 List **two** factors that contribute to a product's carbon footprint. (2)
5 Describe in detail how the Internet of Things (IoT) is impacting our daily lives. (3)

# 1.2 Critical evaluation of new and emerging technologies

## 1.2.1 How to critically evaluate new and emerging technologies

REVISED

Table 1.3 Questions a designer might ask when evaluating design proposals

| Budget constraints | What is the budget for the development of the product? How much is available for materials and components? This could mean using cheaper component parts, which could impact on quality. |
|---|---|
| Timescale | What is the timescale for the development of the product? What is needed to ensure it is fit for purpose and tested before launch? Failure to meet a set timescale could impact on future sales or product success. |
| Who the product is for | Who is the target customer? Have user needs and wants been considered? Meeting the needs and wants of users is critical for a successful outcome. |
| Materials used | What materials support the function of the product best? How can emerging technologies improve the product? Emerging technologies can lead to competitive advantages. |
| Manufacturing capabilities | What are the most efficient methods of manufacture? Can emerging technology support manufacture? 3D printing (**additive manufacture**) is an emerging technology that offers additional possibilities. |

> **Additive manufacture**: a system of printing layer by layer.

# 1.2.2 How critical evaluations can be used to inform design decisions

**Table 1.4** Applying critical thinking to contemporary problems and potential future scenarios

| | |
|---|---|
| **Natural disasters** | CAD programs allow for 3D simulations of real-life disasters and help visualise how a structure might respond. |
| **Medical advances** | Emerging technology has had a major impact on the medical profession, for example the 3D printing of **synthetic** cartilage fits the knee joint perfectly, and the use of **Nitinol** (an **alloy** of nickel and titanium) in medical implants such as stents. |
| **Travel** | Electric cars are seen by many as the way forward. Battery technology continues to evolve and driving range will increase. Electric cars do not produce emissions, are efficient and environmentally friendly. |
| **Global warming** | The main cause of global warming is the release of greenhouse gases such as carbon dioxide into the atmosphere. Solar and wind farms are examples of technology that can help reduce these emissions. |
| **Communication** | Emerging technology in fibre optics and wireless (Bluetooth) have led to a vast range of digital devices that allow easy communication on a global scale. |

**Synthetic**: human-made.

**Nitinol**: an alloy of nickel and titanium.

**Alloy**: a metal mixed with another metal or element (such as carbon) to improve its properties in some way.

# 1.2.3 Ethical perspectives

**Where was it made?**
In some countries, working conditions are not considered safe. The textile industry has many examples of poor working practices.

**Who was it made by?**
Some countries still use child labour despite efforts to stop it. Are the workers fully trained and supported in the jobs they do?

**Who will benefit?**
If workers are well paid then they and their families benefit. If not, it is the manufacturer or the people who buy cheap products who benefit.

**Figure 1.3** Ethical questions

## Fairtrade

Fairtrade develops fair trading relationships between producers, businesses and consumers.

Social, economic and environmental Fairtrade Standards are set for all organisations involved in the supply chain.

Workers receive fair wages and also a Fairtrade Premium to invest in their business and community projects like health care and education.

**Figure 1.4** The FAIRTRADE Marks

Conditions must be of a satisfactory standard to guard against **exploitation** so that workers' rights are protected.

When consumers choose products that carry the FAIRTRADE Marks, they are supporting disadvantaged workers and producers in developing countries.

**Exploited**: when someone is unfairly taken advantage of.

## 1.2.4 Environmental perspectives

REVISED

It is important that designers:

- choose materials that are environmentally friendly
- use materials from sustainable sources
- consider using less packaging or recycled packaging
- consider what happens once products are no longer needed, making recycling easier by ensuring that materials can be separated easily and reused.

**Life-cycle analysis (LCA)** looks at the environmental impact of a product throughout its entire life. In LCA, the following factors should be considered: source of raw materials, material processing, manufacturing, use, end of life and transportation, as well as the energy used at various points during its life cycle.

**Life-cycle analysis (LCA)**: analysing the impact a product has on the environment during its entire life cycle.

### Energy used and consumption during manufacture and transportation

- Transportation of products uses energy from fossil fuels, increasing a product's carbon footprint (see Section 1.1.3 Sustainability on page 8). Reducing transportation by using local manufacturers with locally sourced materials can reduce the carbon footprint.
- **Environmental directives** (laws) from the EU or organisations such as World Energy Council are targets for governments to work towards to reduce energy consumption and pollution, and to eliminate the disposal of hazardous waste into the environment. These directives also cover climate change, air pollution and the protection of wildlife.

**Typical mistake**

A life-cycle analysis is often confused with a product life cycle, which refers to the sales it achieves.

**Environmental directives**: laws that aim to protect the environment.

**Exam tip**

Questions beginning with 'evaluate' or 'analyse' require an extended answer where detail is needed to gain full marks. 'Analyse' requires reasoning in the answer, while 'evaluate' needs evidence of appraisal.

### Now test yourself

TESTED

1 Explain why a limited budget could impact on the design of a new product. (3)
2 Describe **one** example where emerging technology is supporting the medical profession. (1)
3 List **four** factors a designer would look at when carrying out a life-cycle analysis (LCA). (4)
4 Explain what is meant by 'additive manufacture'. (2)

# 1.3 How energy is generated and stored

## 1.3.1 Sources, generation and storage of energy

**Table 1.5 Non-renewable energy sources**

| Oil | Crude oil is extracted from the Earth and refined into liquid fuels such as petrol. It can also be used to generate electricity in power stations. |
|---|---|
| Gas | Extracted through drilling and piped through the national grid to houses and factories. It can also be used to generate electricity in power stations. |
| Coal | Mined from the ground and burnt in power stations to generate electricity. |

Issues surrounding the use of fossil fuels:

- Fossil fuels have high energy density – they hold a lot of chemical energy per kilogram of fuel – making them ideal for transportation.
- Gases such as carbon dioxide and pollutants such as sulphur dioxide are emitted when fossil fuels are burnt. This can cause breathing problems and contributes to global warming.
- Fossil fuels cannot be replaced and will eventually run out.

**Table 1.6 Renewable energy sources**

| Wind | A wind turbine extracts energy from the wind. The blades are connected to a generator that produces electricity. |
|---|---|
| Solar | Photovoltaic (PV) panels produce electricity when exposed to the sunlight. |
| Hydroelectric | Dams house large turbines that trap water. When the water is released, the pressure turns the turbines and generate electricity. |
| Biofuels – biodiesel and biomass | Wood not used in the timber industry is chipped and used as fuel (biodiesel) instead of burning coal. In some biomass schemes plants such as soy are grown to produce materials that can be processed into biofuels. |
| Tidal | Energy is extracted from the rise and fall of the tide. Large turbines can be placed in areas where there is high tidal movement. |

**Table 1.7 Advantages and disadvantages of renewable energy sources**

| Advantages | Disadvantages |
|---|---|
| Renewable energy sources are non-polluting and considered better for the environment. | Initial outlay for the equipment needed for renewable energy is expensive (it produces free energy following installation, however). |
| Although biomass fuels release carbon dioxide as they are burned, trees are replanted, which absorb carbon dioxide as they grow. The process is classed as **carbon neutral**. | Wind and solar power depend on weather conditions and therefore cannot be relied on. |
| Manufacturers are installing equipment to recover waste energy to heat their offices. This will reduce energy bills, but is also a more ethical approach to construction. | In order to build dams to generate hydroelectric power, valleys in rural areas must be flooded. This can damage the natural habitat for wildlife. |

**Carbon neutral**: no net release of carbon dioxide.

## 1.3.2 Powering systems

- Mains electricity is delivered to homes and businesses via the National Grid using underground or overhead cables via pylons.
- Energy from solar and wind farms also contributes to the National Grid.
- Batteries store electrical power. Battery capacity varies depending on the number of cells and the voltage within each cell.
- Battery technology continues to evolve – the size of batteries has significantly been reduced but the capacity to power a device has increased.

## 1.3.3 Choosing appropriate energy sources

Using renewable energy instead of burning fossil fuels is seen by many as the cleaner and greener way forward. However, there is still an environmental cost:

- Many people consider wind farms and solar panels unsightly and noisy.
- There are concerns that birds could be harmed by wind turbine blades.

Compact renewable energy sources have been developed to be used on portable devices:

- Small solar PV panels can produce a small current to recharge a battery. Flexible solar PV panels can be found on clothing and bags, which can charge a mobile phone.
- Low-powered products can be charged from a small wind generator.
- Clockwork wind-up mechanisms can provide a temporary source of power for mechanical or electronic products.
- Technology allows for electrical devices to be far more efficient in terms of energy used. For example, using **light-emitting diode (LED)** lights instead of filament lamps reduces energy consumption.
- Domestic appliances carry an energy rating label. A+++ is the most efficient; while G is the least.
- Standard connections allow products to be sold worldwide, for example USB slots on computers.
- Rechargeable products are often seen as an advantage by consumers.

> **Light-emitting diode (LED):** an output device that produces light when current flows from anode to cathode; they are more energy efficient than traditional filament lightbulbs.

**Figure 1.5** Electronic road sign powered by solar energy

### Now test yourself

1. Biomass fuels are said to be carbon neutral. Explain the meaning of this term. (2)
2. Describe in detail **one** advantage and **one** disadvantage of wind power. (2)
3. Discuss the disadvantages of using fossil fuels to provide energy. (4)
4. State the correct name for the cells found on solar panels. (1)
5. Explain how technology has impacted the development of batteries. (2)

> **Exam tip**
>
> Make sure you fully understand the meaning of key words and terminology. This will help you to understand the context of each question and enable you to write a full and detailed answer.

> **Typical mistake**
>
> Read exam questions carefully and make sure you fully understand what the question is asking before attempting an answer. The main point of the question can often be missed if an answer is rushed.

# 1.4 Modern and smart materials, composite materials and technical textiles

## 1.4.1 Modern and smart materials

Smart materials change or react to a change in their environment, such as temperature, light, pressure or electrical input. Reactions include a change in colour, shape or resistance.

### Shape-memory alloys

- Shape-memory alloys (SMAs) appear to have a memory that allows them to return to an original shape if heated.
- Nitinol, an alloy of nickel and titanium, is a commonly used SMA. Possible uses include medical applications such as medical fastenings used in bone fractures.

### Nanomaterials

- Nanomaterials are defined as having particles on nanoscale dimensions, approximately equivalent to one billionth of a metre.
- In textiles, nanoparticles can alter the properties of a fabric, for example adding a stain-resistant finish.
- New materials have been developed as a result of miniaturisation of component parts, which has led to the development of **conductive fibres**, commonly referred to as **e-textiles**.
- Conductive fibres and threads developed from carbon, steel and silver can be woven into textile fabrics and made into clothing, or the conductive threads can be sewn into a product to connect a circuit. This allows interaction between the fabric and the user.

### Reactive glass

- Reactive glass reacts almost immediately to bright sunlight. It is controlled by a **light-dependent resistor (LDR)**.
- Smart glass reacts in a similar way but has to be activated by a switch. It is opaque when switched off. It is used in homes to give privacy instead of using curtains or blinds.

### Piezoelectric materials

- Piezoelectric materials create an electrical charge when exposed to compression or pressure. Possible uses include sound or sonar detection and lighters for grills and barbeques.

### Temperature-responsive polymers

- Photochromic pigments or dyes change colour in response to changes in light. For example, sunglasses can change colour in response to UV radiation.
- Thermochromic pigments or dyes change colour in response to a change in heat and can be engineered to specific heat ranges.

> **Conductive fibres**: fibres that conduct electricity.
>
> **E-textiles**: textile fabrics that have integrated and interactive devices built into the fabric that react with the user.
>
> **Light-dependent resistor (LDR)**: a sensor input that can detect changes in light levels.

**Figure 1.6 Flexible solar panels can now be integrated into textiles fabrics**

**Figure 1.7 Thermochromic mugs change colour when boiling water is poured in**

Answers, glossary and quick quizzes at www.hoddereducation.co.uk/myrevisionnotes

## Conductive inks

Conductive inks contain conductive particles that allow designers to hand-draw or digitally print circuits that are fully functional.

# 1.4.2 Composites

**Composites** are created when two or more materials are joined together to create a new, enhanced material. One material is known as the **matrix** while the other is the **reinforcement**.

> **Composite material**: a new material created by combining two or more different types of material.
>
> **Compressive strength**: the ability to withstand a pushing (squashing) force.

## Concrete

- Concrete is made up of various aggregates (gravel, rock or sand), cement and water. The cement binds the materials together when water is added. Concrete has a high **compressive strength** but is much weaker under tension.
- Uses include the construction of buildings and roads, as well as smaller projects such as kerbs and drainage.

## Plywood

- Plywood is made up of several layers of wood glued on top of each other. Each layer is laid at a 90-degree angle to the previous layer, giving plywood a consistent strength. The outside layers might be a more expensive wood such as birch.
- Uses include roofing, flooring and furniture.

**Figure 1.8 Plywood**

## Fibre/carbon/glass and reinforced polymers

- Carbon fibre reinforced polymer (CFRP) is a composite consisting of woven carbon fibre strands encased in a polymer resin.
- Carbon fibre strands have a high tensile strength and the polymer resin is lightweight and rigid; when combined, this creates a high-performance engineered material. Possible uses include high-performance sports equipment.
- Glass reinforced plastic (GRP) is a composite of glass fibres and a polyester resin. Glass fibres create rigidity and the resin makes GRP tough and lightweight.
- GRP is difficult to recycle because the process of combining the glass fibres and resin cannot be reversed.

**Figure 1.9 CFRP matting**

## Robotic materials

- Scientists are developing materials that will appear to 'think' for themselves. It is anticipated that these materials will sense something, process it, then react to that sensation, although this is some way off from being a reality.
- Possible uses could include camouflage or load balancing.

# 1.4.3 Technical textiles

- Technical textiles are engineered with specific performance characteristics that suit a particular purpose or function.
- Technical textiles can be woven or bonded, synthetic or natural.

**Table 1.8 Technical textiles and their uses**

| Textile | Where used | Uses |
|---|---|---|
| Agro-textiles | Agricultural industry | Mainly used in crop protection to suppress weeds or for moisture control |
| Construction textiles | Construction industry | Netting used on scaffolding to protect passers-by from falling debris; linings for swimming pools and ponds |
| Geotextiles | Civil engineering, road construction and building | Permeable fabrics used with soil that have the ability to filter, separate, protect and drain; control of embankments on the sides of roads and protection to prevent coastal erosion |

## Domestic textiles

- Domestic textiles have various applications in the home, from padding inside soft furnishing to cleaning cloths.
- **Microfibres** are engineered fibres that are about 100 times smaller than a human hair. When used in cleaning clothes they are highly effective in attracting dust.

> **Microfibres**: tiny fibres about 100 times thinner than a human hair.

## Environmentally friendly textiles

- Environmentally friendly textiles come from sustainable sources such as plants and animals.
- However, to be considered truly eco-friendly they have to be grown organically, without the use of pesticides and fertilisers to boost crops.

## Protective textiles

- Gore-Tex is the most well-known breathable fabric and consists of three or more fabrics laminated together with a breathable **hydrophilic membrane** in the middle:
  - Warm air and tiny droplets of moisture from perspiration can permeate out through the breathable membrane but moisture from larger rain droplets and wind cannot enter.
  - It is used in high performance clothing and footwear, and it helps to regulate body temperature.
- **Aramid fibres** are engineered fibres with very high tensile strength and resistance to heat.

> **Hydrophilic membrane**: a solid structure that stops water passing through but, at the same time, can absorb and diffuse fine water vapour molecules.
>
> **Aramid fibres**: engineered synthetic polymers that are very strong and resistant to heat.

> **Exam tip**
>
> Be able to fully explain how smart, composite and technical textiles are used in the design and manufacture of products. Know their specific properties and what makes each suitable for a particular purpose in a product.

> **Typical mistake**
>
> Don't confuse technical textiles such as Gore-Tex with fabrics such as cotton that absorb moisture making them appear breathable.

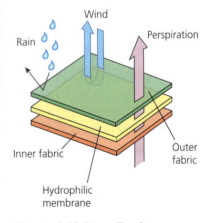

**Figure 1.10 Gore-Tex is an example of a laminated fabric**

**Table 1.9** Aramid fibres

| Fibre | Properties | Uses |
|---|---|---|
| Kevlar® | Lightweight, flexible and extremely durable aramid; excellent resistance to heat, corrosion and damage from chemicals and high tensile strength-to-weight ratio | Used in protective clothing such as police body armour, where the fibre is woven in a lattice that provides protection against knife attack |
| Nomex® | Extremely strong synthetic fabric that can withstand exposure to the most extreme conditions | Primarily used where resistance to heat and flames is essential, for example firefighters' uniforms |

## Sports textiles

- **Micro-encapsulation** is a process of applying microscopic capsules to fibres and fabrics.
- The capsules can contain vitamins, therapeutic oils, moisturisers, antiseptics and antibacterial chemicals that are released through friction.
- Rhovyl® is a non-flammable, synthetic fibre that is crease resistant, has good thermal and acoustic properties, is antibacterial and comfortable to wear.
- The construction of the fibre gives fabrics the ability to wick away moisture, such as perspiration, through the fabric. It also dries quickly, meaning it does not retain odours, making it ideal for socks and sportswear.

> **Micro-encapsulation**: tiny microscopic droplets containing various substances applied to fibres, yarns and materials including paper and card.

### Now test yourself

TESTED ☐

1  Describe what gives plywood its strength. (2)
2  Explain why carbon fibre reinforced polymer (CFRP) is a suitable material for a racing bike. (3)
3  Explain how micro-encapsulation is beneficial to a patient when used in medical dressings. (2)
4  State what is meant by 'smart glass'. (1)
5  Describe **two** reasons why the aramid fibre Nomex® is used in protective clothing. (2)

# 1.5 Mechanical devices

## 1.5.1 Types of movement

REVISED ☐

The four main types of movement are:

- **linear**: movement in a straight line
- **reciprocation**: movement that goes back and forth in a straight line
- **rotary**: movement that turns in a circle
- **oscillation**: a swinging movement from side to side.

# 1.5.2 Classification of levers

- **Levers** consist of a rod pivoted on a **fulcrum**.
- In a class 1 lever, the effort and the load are on opposite sides of the fulcrum. An example is a see-saw.
- In a class 2 lever, the load and the effort are on the same side of the fulcrum, with the load nearest the fulcrum. The effort required is less than the load. An example is a wheelbarrow.
- In a class 3 lever, the load and effort are on the same side of the fulcrum, but this time the effort is nearest to it. The effort required is greater than the load, but the load can be moved faster. An example is a pair of tweezers.

> **Lever**: a rod pivoted on a fulcrum – the class of lever (1, 2 or 3) affects how much effort is needed to move the load.
>
> **Fulcrum**: the point, or pivot, on which a lever sits.

> **Exam tip**
>
> Make sure that you show all of your working when performing calculations related to mechanical systems, including writing down the formula.

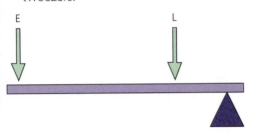

**Figure 1.11** A class 2 lever

$$\text{Mechanical advantage of a lever} = \frac{\text{Load} \left( F_b \right)}{\text{Effort} \left( F_a \right)}$$

$$\text{Velocity ratio of a lever} = \frac{\text{Distance moved by effort}}{\text{Distance moved by load}}$$

$$\text{Efficiency of a lever} = \frac{\text{Work done on the load}}{\text{Work done by the effort}} \times 100$$

# 1.5.3 Linkages

- **Linkages** are used to change the size of a force, the direction of motion and/or the type of motion.
- In a **reverse motion linkage**, the bottom link moves to the left as the top link moves to the right. If the distance between the fixed pivot and the moving pivots is equal, then the output force is the same as the input force. The output force can be increased or decreased by changing the position of the fixed pivot.
- In a **bell crank** the output movement is at 90 degrees to the input movement. The output force is greater than the input force when the fixed pivot is closer to the output lever.

> **Linkage**: levers connected together via fixed and moving pivots; they are used to change the size of a force and/or the direction of motion.

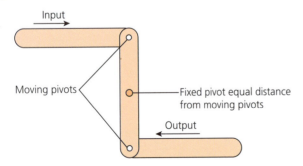

**Figure 1.12** A reverse motion linkage

# 1.5.4 Cams

- **Cams and followers** turn rotary motion into reciprocating motion. The follower moves up and down as the cam rotates.
- With a **pear-shaped cam**, the follower remains stationary (dwells) for half of the cycle. It is then pushed up as the point of the cam approaches. Finally, as the point passes, the follower falls and then dwells again.
- **Eccentric cams** are circular but have an off-centre rotating shaft. The rise and fall produced is symmetrical.
- With a **snail/drop cam** the follower dwells for the first 120 degrees or so of the cycle. It then rises slowly before dropping suddenly when it reaches the peak of the snail shape.

> **Cam and follower**: a mechanism that converts rotary motion into reciprocating motion.

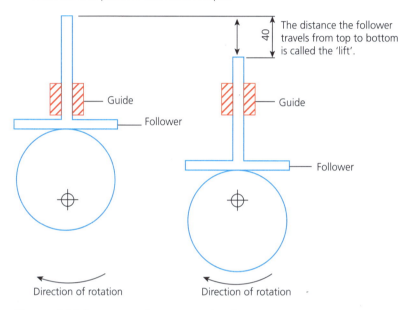

The distance the follower travels from top to bottom is called the 'lift'.

**Figure 1.13** An eccentric cam mechanism

# 1.5.5 Followers

- **Knife followers** have a sharp point that makes contact with the cam. They often wear quickly, leading to the mechanism not working correctly.
- **Roller followers** replace the sliding motion between the cam and follower with a rolling motion. This reduces wear.
- **Flat followers** are used where space is limited.

# 1.5.6 Pulleys and belts

- V-belt pulley systems transmit rotary motion and torque. A larger pulley driving a smaller pulley results in increased speed but decreased torque. A smaller pulley driving a larger pulley achieves the opposite.
- The speed of a pulley is measured in revolutions per minute (rpm).

> **Pulley and belt**: A system that uses pulley wheels and a belt to transmit rotary motion and torque.

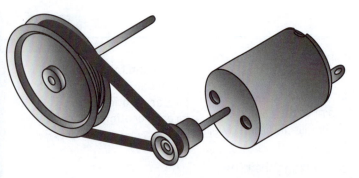

**Figure 1.14** An electric motor driving a simple pulley system

$$\text{Velocity ratio (VR) of a pulley system} = \frac{\text{Diameter of the driven pulley}}{\text{Diameter of the driver pulley}}$$

$$\text{Speed of the driven pulley (rpm)} = \frac{\text{rpm of the driver pulley}}{\text{Velocity ratio}}$$

## 1.5.7 Cranks and sliders

REVISED

- **Crank and slider** mechanisms convert rotary motion to reciprocating motion, and vice versa. To create reciprocating motion, the crank rotates and the connecting rod pushes the slider backwards and forwards.
- Piston engines in cars are an example of the reverse.

**Crank and slider:** a mechanism that converts rotary motion to reciprocating motion and vice versa.

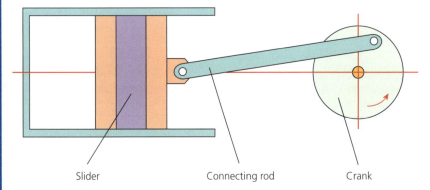

Slider          Connecting rod          Crank

**Figure 1.15** A crank and slider mechanism

## 1.5.8 Gear types

REVISED

- In a simple two **gear train**, a larger gear driving a smaller gear results in increased speed but decreased torque. A smaller gear driving a larger gear achieves the opposite.

$$\text{rpm of the driven gear} = \frac{\text{rpm of the driver gear}}{\text{Gear ratio}}$$

**Gear train:** a collection of spur gears that transmit rotary motion and torque.

- **Compound gear trains** are created when one or more of the shafts holds two gears. They are used when large increases in speed or torque are needed and when space is limited.
- An **idler gear** is placed between two gears to ensure that they have the same direction of motion.
- **Bevel gears** are used to change the direction of a shaft's rotation, usually through 90 degrees.
- **Rack and pinions** change rotary motion to linear motion. The teeth on the circular pinion gear mesh with those on the rack. As the pinion gear turns, the rack moves in a straight line.

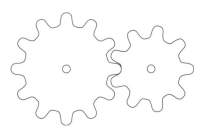

**Figure 1.16** A simple gear train

**Typical mistake**

It is easy to mix up the driven pulley and the driver pulley when calculating velocity ratios. Ensure you know which one is which.

Answers, glossary and quick quizzes at www.hoddereducation.co.uk/myrevisionnotes

# 1.6 Electronic systems

## 1.6.1 Sensors

REVISED

- Sensors are input devices. They take signals from the environment around them, such as light and temperature levels, and turn them into electronic signals, such as current, voltage or resistance.
- **LDRs** change the light level that is detected into a resistance. The resistance decreases as the brightness increases. LDRs are often used in outdoor street lamps and night lighting.
- **Thermistors** change the temperature level that is detected into a resistance. Thermistors are often used where it is important to know the temperature, such as inside a greenhouse or a refrigerator.

**Figure 1.17 A light-dependent resistor (LDR)**

> **Thermistor**: a sensor input that can detect changes in temperature levels.

## 1.6.2 Control devices and components

REVISED

- **Switches** are used to either 'make' (allow current to flow through) or 'break' (not allow current to flow through) a circuit. An example is a light switch that turns a lamp 'on'.
- Switches are described by either the number of poles and throws that they have, or by their function. For example, a push to make switch 'makes' a circuit when it is pressed.
- Other examples of switches include rocker switches, toggle switches, tilt switches, reed (magnetic) switches and micro switches.
- **Transistors** are used as either an electronic switch or a current amplifier. A transistor has three leads called the base, collector and emitter. When the voltage at the base reaches approximately 0.6 V, current is allowed to flow between the collector and the emitter.
- **Resistors** are used to reduce the flow of current in a circuit. The higher the value of a resistor, the more resistance it has. The amount of resistance is measured in ohms, or $\Omega$.

**Figure 1.18 A push to make switch**

> **Switch**: a component that can be used to either 'make' or 'break' a circuit.
>
> **Transistor**: a component that acts as either an electronic switch or a current amplifier.
>
> **Resistor**: a component that reduces the flow of current.

## 1.6.3 Outputs

REVISED

- Outputs take electronic signals and turn them back into 'real world' signals, such as light or sound.
- **Buzzers** use an internal oscillator to produce sounds at different frequencies when current flows through them. They are often used in alarms, warning systems and electronic children's toys.

> **Buzzer**: an output device that produces sound when current flows through it.

- **LEDs** produce light when current flows from the anode to the cathode. They can be damaged by too much electrical current, so they usually require a protective resistor connecting in series with them.
- LEDs use far less current and thus save energy. They also last much longer as they do not 'burn out'.

**Typical mistake**

Misunderstanding the difference between input, control and output components and their main functions in a system.

**Now test yourself**  TESTED ☐

1 A designer is developing an idea for an outdoor security light. The light must come on automatically when it is dark. Name a suitable input sensor and output device for this system. (2)
2 Describe how a transistor operates when being used as an electronic switch. (3)
3 State the main function of a resistor in a circuit. (1)
4 Give **two** applications of buzzers in electronic systems. (2)

**Exam tip**

Make sure you understand the working characteristics, applications, advantages and disadvantages of the main input, control and output components.

# 1.7 The use of programmable components

## 1.7.1 How to make use of flowcharts  REVISED ☐

- Programmable components are used to replace analogue integrated circuits. They can be programmed to perform tasks such as responding to sensors, producing time delays and counting.
- An example is a microcontroller, which has several pins for the connection of different input and output devices.
- Flowchart software is commonly used to program microcontrollers. Standard symbols are joined together using arrows to create a set of instructions for how the system will operate. This is then downloaded on to the chip. Each symbol represents a type of command that can be edited for a specific purpose within the program.
- This type of programming is highly visual, so it is easier to make changes or spot errors. Complex programming knowledge is not necessary. Programs can be simulated on-screen to check that they will work.

**Typical mistake**

Not using the correct flowchart symbols for each programming command – make sure you know when each of these should be used in a program.

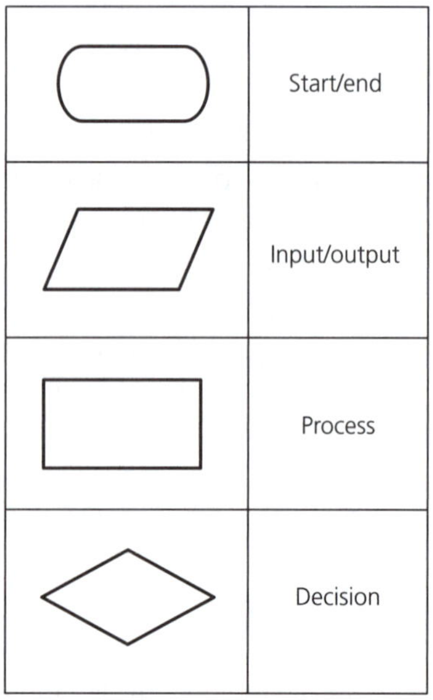

## 1.7.2 How to switch outputs on/off  REVISED ☐

- In most flowchart software, an output is turned on by inserting the word 'high', followed by the pin number that the output is connected to, into the **input/output** command symbol.
- For example, 'High 2' would turn on the output connected to pin 2. To turn the output off, the word 'low' is used instead.
- **Decision** commands are used to check the state of a digital input device, such as a switch. This is achieved by inserting a question into

| | |
|---|---|
| (rounded rectangle) | Start/end |
| (parallelogram) | Input/output |
| (rectangle) | Process |
| (diamond) | Decision |

**Figure 1.19 Common flowchart symbols**

the decision command symbol, for example 'is pin 3 on?'. If the answer is yes, the program follows the 'yes' (Y) arrow; if it is no, it follows the 'no' (N) arrow.

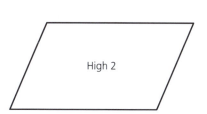

**Figure 1.20 A command that would turn an output device connected to pin 2 'on'**

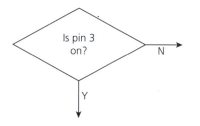

**Figure 1.21 A decision command checking whether a digital input device connected to pin 3 is 'on'**

> **Start/End**: a command that marks the start or end of a program.
>
> **Input/output**: a command that is used to change the state of an input or output pin.
>
> **Decision**: a command that checks the state of a digital input device.

### 1.7.3 How to process and respond to analogue inputs

`REVISED` ☐

Many sensors, such as LDRs and thermistors, produce analogue signals. A special type of decision command, called **compare**, can be used to check whether an analogue signal is greater or less than a particular value, known as the threshold.

> **Compare**: a command that checks whether an analogue value is greater or less than the threshold.

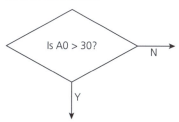

**Figure 1.22 A compare command checking whether an analogue input signal is greater than 30**

### 1.7.4 How to use simple routines to control outputs with delays, loops and counts

`REVISED` ☐

- One of the main reasons for using programmable components is that complex **process** commands can be created with less circuitry, as programming replaces the hardware.
- To create time delays the **wait** command can be used. For example, the program in Figure 1.23 will turn on an output connected to pin 2 when pin 3 detects that a switch has been pressed. The output will remain on for one second and then turn off. The program will then wait another second before looping back to repeat, creating a continuous pulsing sequence.
- Most flowchart programming software has a type of process command that can **count** the number of times an input pin goes from low to high within a certain time period, for example how many times a switch is pressed.

> **Process**: a command that can be used to perform processing functions such as counting or creating time delays.
>
> **Wait**: a command that is used to create time delays.
>
> **Count**: a command that can count the number of times an input pin goes from low to high.

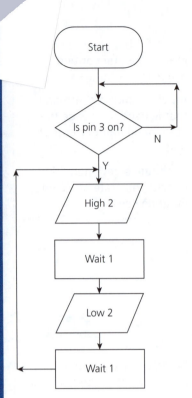

**Figure 1.23** A program that uses time delays and
loops to create a continuous pulsing sequence

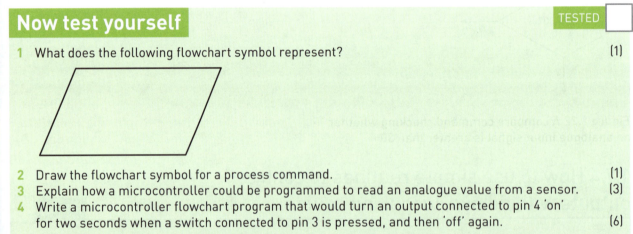

## Now test yourself

TESTED

1  What does the following flowchart symbol represent? (1)

2  Draw the flowchart symbol for a process command. (1)
3  Explain how a microcontroller could be programmed to read an analogue value from a sensor. (3)
4  Write a microcontroller flowchart program that would turn an output connected to pin 4 'on'
   for two seconds when a switch connected to pin 3 is pressed, and then 'off' again. (6)

# 1.8 Ferrous and non-ferrous metals

## 1.8.1 Ferrous metals

REVISED

- **Ferrous metals** contain iron so they corrode quickly.
- They are magnetic and corrode quickly if not treated with a suitable
  surface finish.

**Ferrous metals**: metals that
contain iron.

### Mild steel

- Mild steel is also known as low-carbon steel because it has the lowest
  carbon content of the different steel types.
- Mild steel contains around 99 per cent iron and a maximum of 0.25 per
  cent carbon.

Answers, glossary and quick quizzes at www.hoddereducation.co.uk/myrevisionnotes

- Other elements such as manganese, silicon and sulphur make up the remaining fraction.
- Mild steel is relatively affordable compared to other types of steel.
- It is easy to machine and weld.
- Common uses are structural steel in buildings, signs, car and vehicle bodies and panels, and wire.

## Stainless steel

- Stainless steel contains iron and between 15 and 30 per cent of other metals (chromium and nickel).
- It has a high resistance to corrosion compared to other types of steel.
- It is tough and ductile with high strength and temperature resistance.
- Common uses are medical instruments, kitchen utensils and appliances, vehicle exhaust systems, hardware and structural/architectural items.

## Cast iron

- Cast iron is a mixture of iron and two per cent or more of carbon.
- It is heavy and has good compressive strength but is quite **brittle** and can crack or shatter under impact.
- It has a low melting point and is more resistant to wear and corrosion than mild steel.
- Common uses are pipework, engines, cooking utensils, and garden ornaments and furniture.

**Brittleness:** hard but easily broken or shattered.

## 1.8.2 Non-ferrous metals

REVISED

- **Non-ferrous metals** do not contain iron.
- They are much more corrosion resistant than ferrous metals.
- They are generally more expensive than ferrous metals.

**Non-ferrous metals:** metals that do not contain iron.

## Aluminium

- Aluminium has a high strength-to-weight ratio as it is lightweight.
- It has a shiny, attractive appearance that requires no additional finish.
- It is easy to cut, shape and weld.
- Common uses are drinks cans, foils, window frames and aeroplane parts.

## Copper

- Copper is an excellent electrical and thermal conductor that does not tarnish or corrode easily.
- It is tough but ductile, so it can be easily shaped or drawn into a wire.
- Copper can be joined easily by soldering or brazing.
- It can be polished to an attractive appearance and golden colour.
- Common uses are water pipes, electrical cables, jewellery, statues and outdoor decorative features.

### Brass

- Brass is an alloy of copper and zinc.
- Alloys are metals made by combining two or more metals and other elements together to improve their properties.
- Brass has good resistance to wear and corrosion, and has an attractive appearance.
- Common uses for brass are nuts, bolts and screws, door furniture and bathroom fittings.

**Figure 1.24 Brass is an alloy made from copper and zinc**

**Typical mistake**

Unlike gold and silver which are non-ferrous metals, brass is an alloy and not a 'pure' metal.

## 1.8.3 Properties

REVISED

The properties of different metals are the main factors that will influence their selection and use. The general properties of metals are:

- **high melting point**: resistant to very high temperatures
- **finish**: can be polished to an attractive shine
- **conductive**: allows electricity and heat to pass through easily
- **malleable**: can be hammered and bent into shape without breaking
- **ductile**: can be stretched into a thin wire without becoming weaker or brittle in the process
- **hardness**: resistant to impact and abrasion.

**Conductive**: allows electricity and heat to pass through easily.

**Malleable**: can be hammered and bent into shape without breaking.

**Ductile**: can be stretched into a thin wire without becoming weaker or brittle in the process.

**Now test yourself**

TESTED

1 Name the **two** differences between ferrous and non-ferrous metals. (2)
2 What is an alloy? (2)
3 All metals are good conductors. What does this mean? (2)

# 1.9 Papers and boards

## 1.9.1 Paper

REVISED

- Papers and boards come in a wide range of different thicknesses, sizes and types.
- The thickness of paper is known as its 'weight' and is measured in **grams per square metre (gsm)**.

**Grams per square metre (gsm)**: the weight in grams of one square metre of paper.

### Copier paper

- Copier paper is used extensively for printing, photocopying and general office purposes.
- It is approximately 80 gsm.
- It has a smooth surface making it ideal for most printers and photocopiers.

**Exam tip**

Papers heavier than 170 gsm are classified as boards.

## Cartridge paper

- Cartridge paper is available in different weights between 80 and 140 gsm.
- It is thicker and more expensive than layout and copier paper.
- It has a slightly textured surface and is slightly creamier in colour.
- Cartridge paper is used by artists for sketching, drawing and painting as it has an ideal surface for pencil, crayons, pastels, watercolour paints, inks and gouache.

## Tracing paper

- Tracing paper is used for making copies of drawings and fine details.
- It is thin and very transparent.
- It is also hardwearing and strong despite its lack of thickness.
- Mistakes made in pen can be scratched off using a sharp blade.

# 1.9.2 Board

REVISED

- Boards are usually classified by thickness as well as by weight.
- The thickness of board is measured in **microns**; a micron is one-thousandth of a millimetre.

## Folding boxboard

- Folding boxboard is made up of three layers of card.
- The card layers have different qualities and properties: the two outer layers are high quality with a very smooth surface; the middle layer is lower quality but has high bulk and **stiffness**.
- A chemical coating on the top outer layer makes it ideal for high-quality printing and foiling.
- Folding boxboard is used for health and beauty products, frozen and chilled foods, pharmaceutical products and confectionaries.

> **Stiffness**: resistance to bending and flexing.

## Corrugated cardboard

- Corrugated cardboard is a strong but lightweight type of card.
- It is made from two layers of card with a fluted sheet in between.
- The fluted construction makes it stiff and difficult to bend or fold, especially across the flutes.
- Corrugated card can absorb knocks and bumps, and has good heat-insulating properties.
- Common uses are packaging fragile or delicate items, and takeaway foods, for example pizza boxes.

## Solid white board

- Solid white board (also known as mounting board) is a rigid type of card with a thickness of around 1.4 mm (1400 microns) and a smooth surface.
- It is available in different colours but white and black are the most commonly available.
- Mounting board is often used for picture framing mounts and architectural modelling.

### 1.9.3 Properties

- **Flexibility** is the paper's ability to withstand multiple folds before it breaks. This is important in printed products that may be folded and unfolded numerous times during use, such as books, maps and cartons.
- Paper made with virgin pulp, which has longer fibres, has a higher flexibility than recycled paper.
- **Printability** is how easy the paper can be printed on.
- The smoother and flatter the paper's finish, the easier it is to print on and the higher quality the print finish will be.
- **Biodegradability** means that the paper or board can break down and decompose without harming the environment.
- Normal paper and board will biodegrade quickly. Coatings on products such as coffee cups to make them waterproof significantly affect the length of time it takes the paper and board to decompose.

> **Flexibility**: ability to bend and curve evenly and easily.
>
> **Printability**: how easily a material can be printed on.

> **Typical mistake**
>
> Paper is not generally considered to be a strong material. However, for its weight it is a very strong material.

> **Now test yourself**
>
>
> 1 Name **one** use for tracing paper. (1)
> 2 Explain **two** reasons why corrugated cardboard is ideal for takeaway food packaging. (2)
> 3 Describe **one** use for mounting board. (2)

# 1.10 Thermoforming and thermosetting polymers

## 1.10.1 Thermoforming polymers

- **Thermoforming polymers** soften when heated and can be moulded into shape.
- When the polymer cools it hardens and retains the shape.
- If the polymer is reheated it will try to return to its original shape.
- Thermoforming polymers can be reheated and reshaped numerous times.

> **Thermoforming polymer**: a polymer that can be reheated and reshaped a number of times.

### Acrylic

- Acrylic is a rigid, hard polymer usually in the form of a sheet.
- It is available in clear or a wide range of solid and translucent colours.
- Acrylic is easy to cut, bend, shape and polish.
- It has good **impact resistance** but is easily scratched.
- Common uses include as a substitute for glass, illuminated signs, aircraft canopies and car light clusters.

> **Typical mistake**
>
> Many students give acrylic as a polymer used for vacuum forming. Although it is easy to shape and bend using a strip heater, it is not suitable for vacuum forming.

> **Impact resistance**: resistance to knocks and impacts.

### High-impact polystyrene

- High-impact polystyrene (HIPS) is a thin, flexible and lightweight sheet polymer available in a wide range of colours.
- It is inexpensive and easy to cut, shape, bend and mould.
- HIPS is a popular choice for vacuum forming.
- Common uses for HIPS include disposable plates and cups, model kits, food containers and trays.

## Biodegradable polymers

- Biodegradable polymers, or **biopolymers**, are produced from renewable raw materials such as corn starch, plant sugars and vegetable oil.
- They are environmentally friendly as they are compostable and biodegradable.
- Biopol® is a type of biopolymer. It is lightweight, hard and non-toxic.
- Common uses for biodegradable polymers include medical products such as dissolving thread for stitching wounds, bin bags, bottles and cups, and food packaging.

> **Biopolymer**: a biodegradable polymer made from a plant source, such as corn starch or sugar beet.

## 1.10.2 Thermosetting polymers

REVISED

- **Thermosetting polymers** soften when heated and can be moulded into shape.
- During heating, a chemical change takes place.
- Once the thermosetting polymer has cooled, it hardens and sets in shape.
- Once set, the thermosetting polymer cannot be reheated and reshaped.
- If reheated, the thermosetting polymer will melt and burn.
- Thermosetting polymers cannot be recycled.

> **Thermosetting polymer**: a polymer that can be heated and shaped only once.

> **Exam tip**
> Apart from biopolymers, polymers are not considered to be environmentally friendly materials and their use in modern society is causing considerable damage to the environment.

### Polyester resin

- On its own, polyester resin is hard but very brittle and easy to snap.
- When combined and reinforced with fibreglass, it becomes GRP (glass reinforced plastic).
- GRP is very tough and has high impact resistance. It is also lightweight, waterproof and relatively easy to repair if damaged.
- Common uses for GRP include car body panels, sailing boats, canoes, corrugated roofing sheet, and electrical enclosures.

### Urea formaldehyde

- Urea formaldehyde is a rigid, hard but brittle polymer.
- It is available in a wide range of colours.
- It has excellent heat resistance and is a good electrical **insulator**.
- Common uses for urea formaldehyde include electrical fittings (light switches, plug sockets), domestic appliance parts, laminate sheeting and wood glue.

> **Insulator**: a material that does not conduct heat or electricity easily.

## 1.10.3 Properties

REVISED

The general properties of polymers are:

- **thermal insulator**: does not conduct heat easily
- **electrical insulator**: does not conduct electricity
- **toughness**: resistant to impact and wear.

> **Now test yourself**
> TESTED
>
> 1 Explain why thermoforming polymers are considered to be more environmentally friendly than thermosetting polymers. (2)
> 2 Give **two** reasons why polymers are often used for housing electrical equipment. (2)
> 3 Name a common use for HIPS. (1)

# 1.11 Natural, synthetic, blended and mixed fibres, and woven, non-woven and knitted textiles

- Fibres are very fine, hair-like structures that are spun (or twisted) together to make **yarn**.
- Yarns are woven or knitted together to create textile fabrics.

> **Yarn:** spun threads used for knitting, sewing and weaving.

## 1.11.1 Natural fibres

REVISED

- Natural fibres come from natural sources.
- They are sustainable and biodegradable.

**Table 1.10** Sources, properties and uses of natural fibres

| Fibre | Source | Properties | Uses |
|---|---|---|---|
| Wool | Animal: sheep fleece | Warm, **absorbent**, low flammability, good **elasticity**, crease resistant | Warm outerwear (coats and jackets); knitwear; soft furnishings, carpets and blankets |
| Cotton | Vegetable: cotton plant/ seed pod | Absorbent, strong, cool to wear, hardwearing, creases easily, smooth, easy to care for, flammable, can shrink | Clothing, bedlinen, towels, sewing and knitting threads, soft furnishings |

> **Absorbency:** ability to soak up moisture.
>
> **Elasticity:** ability to stretch and return to shape.

## 1.11.2 Synthetic fibres

REVISED

- Synthetic, or manufactured, fibres are artificial.
- Advantage: they can be manufactured to specific requirements.
- Disadvantage: they are not biodegradable and come from unsustainable sources.

**Table 1.11** Sources, properties and uses of synthetic fibres

| Fibre | Source | Properties | Uses |
|---|---|---|---|
| Polyester | Fossil fuels: coal/ petroleum | Strong when wet and dry, flame resistant, thermoplastic, hardwearing, poor absorbency | A versatile fibre used throughout textiles |
| Acrylic | Fossil fuels: petroleum | Strong except when wet, thermoplastic, burns slowly then melts, poor absorbency | Knitwear and some knitted fabrics, fake fur products including toys, upholstery and carpets |

> **Exam tip**
>
> Look at the labels of different products. Try to identify the fibre used and work out what properties the fibre has that will allow the product to function as intended. The more you look at, the better your understanding of material selection will be.

## 1.11.3 Woven textiles

- Woven fabrics consist of **warp** and **weft** yarns in an under/over configuration.
- Different **weave** creates fabrics with different **textures**, patterns and strength.
- **Plain weave** is the most basic weave. It is stable, strong and gives an even surface on both sides of the fabric. An example is calico.
- **Twill weave** gives a diagonal pattern and adds strength to the fabric. An example is denim.

> **Warp**: yarn following the length of a fabric.
>
> **Weft**: yarn running horizontally across a fabric.
>
> **Weave**: the way yarns are laid in a pattern/formation in the construction of fabric.
>
> **Texture**: the feel and finish of a material.

> **Typical mistake**
>
> Don't confuse warp and weft yarns used in fabric construction. Warp run the length of the fabric and weft run across the width of fabric from left to right.

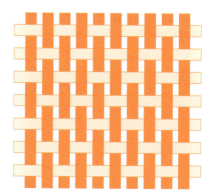

Figure 1.25 **A plain weave structure**

Figure 1.26 **A twill weave structure**

## 1.11.4 Non-woven textiles

### Felted

- Felted fabrics are produced by applying moisture, heat, pressure and friction to a web of fibres, which causes the fibres to bind together. Wool and acrylic fibres are commonly used in this process.
- Felt is used in craft projects, accessories such as hats, and on the surface of snooker tables.

### Bonded

- Non-woven or bonded textiles are **constructed** from a web of fibres, held together by **adhesive** or stitching.
- Most non-woven or bonded fabrics are not strong enough to be made into garments but are often used to reinforce woven and knitted fabrics.
- They are cheap to produce and are often used to make disposable products such as surgical gowns and masks, cleaning wipes and disposable protective suits.

> **Constructed**: the way something has been made.
>
> **Adhesives**: glues.

## 1.11.5 Knitted textiles

- Knitted fabrics are made by creating a series of loops in the yarns that interlock together.
- There are two types of knitted fabrics: **warp knitting** (interlinking vertically) and **weft knitting** (interlinking horizontally).
- Knitted fabric is easy to stretch and warm to wear as the loops trap body heat.

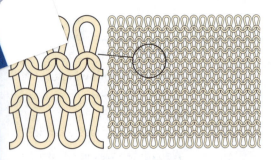

**Figure 1.27 Warp knitting consists of multiple yarns that loop together and interlock vertically**

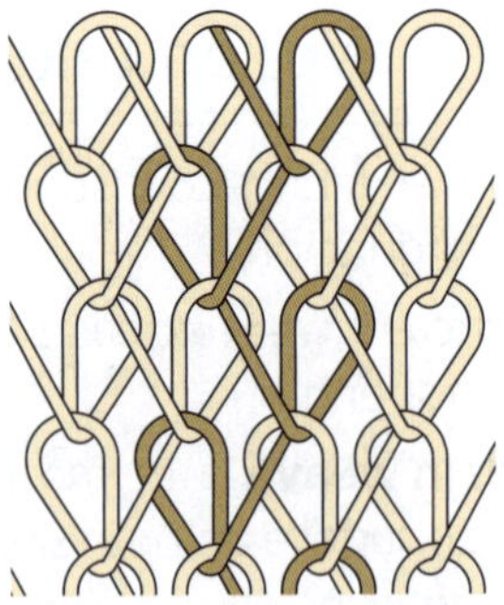

**Figure 1.28 Weft knitting consists of one yarn with a series of loops that interlock horizontally**

## 1.11.6 Properties

REVISED

Fibres have different properties and characteristics depending on the source of fibre and how the fabric has been constructed. This affects what it can be used for. Table 1.12 shows the performance properties of different fibres. The higher the star rating, the greater the property in the stated fibre.

**Table 1.12 Performance properties of fibres**

|  | Wool | Cotton | Polyester | Acrylic |
|---|---|---|---|---|
| **Elasticity** | **** | * | **** | *** |
| **Resilience** | **** | * | **** | *** |
| **Durability** | ** | ** | **** | *** |

### Now test yourself

TESTED

1. Name **one** natural fibre and **one** synthetic fibre. (2)
2. Explain how the type of weave in fabric construction affects the fabric. (2)
3. State **one** characteristic of knitted fabric. (1)
4. Explain why bonded fabrics are often used in disposable products. (2)
5. Describe **two** reasons why natural fibres are considered more environmentally friendly. (2)

# 1.12 Natural and manufactured timbers

- **Natural timber** comes from trees that are cut down into logs and then sawn into planks.
- **Manufactured timber** is made from recycled wood chips and fibres.

> **Natural timber**: timber that comes from trees.
>
> **Manufactured timber**: a human-made board made from recycled timber.

## 1.12.1 Natural timbers – hardwoods

REVISED

- **Hardwoods** come from deciduous trees that shed their leaves every year.
- Hardwood trees grow slowly and can take hundreds of years to grow fully.
- They have thick trunks with branches at the top.
- Hardwoods have a close **grain** and tend to be denser and heavier than softwoods.

> **Hardwood**: timber that comes from deciduous trees.
>
> **Grain**: the longitudinal arrangement of wood fibres.

## Oak

- Oak is a heavy, hard and tough timber.
- It has an open grain and an attractive appearance.
- Oak is good for outdoor use as it is durable and can withstand the elements well compared to other woods.
- Common uses for oak include garden furniture, doors, floors and high-end furniture.

## Mahogany

- Mahogany is a dark coloured close-grained timber.
- It is easy to cut and shape, is available in wide planks and polishes well.
- Common uses for mahogany include furniture, shop fittings, boat building and doors.

## Beech

- Beech is a very tough, hard timber with a close and straight grain.
- It is hardwearing, easy to work and polishes well.
- It has a light brown colour and can be prone to warping.
- Common uses for beech include furniture, toys and wooden tools.

## Balsa

- Balsa is a soft and lightweight timber with a coarse and open grain.
- It is very easy to shape, sand, glue and paint.
- Common uses for balsa include model making, packing cases, surfboards and fishing floats.

> **Typical mistake**
>
> Balsa is a hardwood, despite it being a very soft, lightweight and easy to shape.

## 1.12.2 Natural timbers – softwoods

REVISED

- **Softwoods** come from coniferous (evergreen) trees that have needles instead of leaves.
- Softwood trees keep their needles year-round and grow much faster than hardwood trees.
- They grow tall and straight, and have lots of branches all the way up the trunk.
- Softwood has more knots than hardwood.

> **Softwood**: timber that comes from coniferous trees.

## Pine

- Pine is the most widely available natural timber and is relatively cheap compared to other timbers.
- It has a straight grain and is easy to cut and shape.
- Pine has knots along the grain of the wood. They can weaken its strength but give an attractive appearance when finished.
- Common uses for pine include interior construction work, boxes and crates, flooring and paper (pulpwood).

### Cedar

- Cedar is a lightweight, soft timber with a straight grain.
- It is knot free and very durable.
- Cedar is resistant to rot, weather and insect attack.
- Common uses for cedar include outside joinery, building cladding, bathroom and kitchen furniture, and wall panels.

**Exam tip**

Hardwoods are generally heavier, harder, tougher and more durable than softwoods.

## 1.12.3 Manufactured timbers

REVISED

- Manufactured boards can be made from hardwood and softwood.
- They are made by gluing and compressing wood fibres or layers together.
- They are made in large sheets that are easy to work with.
- Manufactured boards are generally cheaper than 'real' or natural wood.

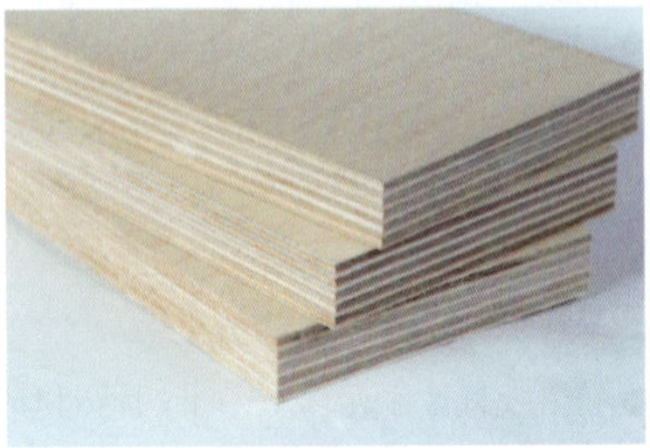

**Figure 1.29** Plywood is made from thin layers of timber placed with alternating grain direction

### Plywood

- Plywood is made by gluing together thin layers of timber with alternating direction of grain.
- It is strong and tough due to its layered construction, with a high strength-to-weight ratio.
- Plywood is easy to cut and shape but can splinter easily.
- It is relatively stable under changes in temperature and moisture.
- Common uses for plywood include structural panelling in building construction and furniture making.

### Medium-density fibreboard

- Medium-density fibreboard (MDF) is made by gluing and compressing wood fibres together into sheets.
- It has a smooth, even surface but chips and damages easily.
- It is easy to cut, shape, paint, glue and stain.
- MDF has poor moisture resistance and degrades quickly when wet.
- Common uses include cheap flat-pack furniture, interior panelling and interior doors.

## 1.12.4 Properties

REVISED

### Hardness

- **Hardness** is the resistance of the wood to denting and wear.
- It varies depending on the direction of the wood grain.
- Hardness of wood is tested by dropping a ball bearing onto it from a set height. This is called the Janka hardness test.

**Hardness**: resistance to impact and abrasion.

### Toughness

- Toughness is the amount that timber can be bent without breaking or splitting.
- Toughness is affected by the timber's internal cell structure.
- Most hardwoods have interlocking cell structures and are generally tougher than softwoods.

## Durability

- **Durability** is the wood's ability to resist decay from exposure to the elements and attack from insects.
- Hardwoods such as teak are the most durable woods and can last for many years when exposed to the elements.
- Some types of bamboo have a durability of only a few months.

**Durability**: resistance to wear, long lasting.

# 1.13 Contexts which inform design outcomes

## 1.13.1 Discriminating between materials, components and manufacturing processes

REVISED

Properties may change depending on things such as temperature or the shape and direction of the material. Other properties will remain the same regardless of the environment or conditions.

Mechanical properties include:

- compressive strength – resistance to a pushing force
- tensile strength – resistance to a pulling force
- durability – resistance to wear, long lasting
- hardness – resistance to abrasion
- stiffness – resistance to bending and flexing
- elasticity – ability to stretch and return to shape
- impact resistance – resistance to knocks and impacts
- brittleness – hard but easily broken or shattered
- water resistance – resistance to water or moisture
- absorbency – ability to soak up moisture.

Manufacturing properties include:

- machinability – ability to be easily milled and drilled
- workability – ability to be placed and compacted
- ductility – ability to stretch, not break under stress or impact
- malleability – ability to be hammered, pressed or rolled without breaking.

Other properties include:

- acoustic – acoustic absorption
- chemical – corrosion resistance
- electrical – conductivity, resistance and so on
- thermal – conductivity, flash point and so on
- magnetic – attractive to magnets.

**Typical mistake**

Malleability and ductility are similar properties and often confused. Ductility relates to the ability of a material to stretch without damage. Malleability relates to the ability of a material to deform under compression.

**Exam tip**

By learning the properties of materials, you can understand why they have been used to make certain products. This will help you to answer questions in the exam about individual products and their fitness for purpose.

## Advantages and disadvantages

- The choice of materials impacts the success of the final product.
- The advantages of a particular material may make it an obvious choice for a product, but the designer must also consider disadvantages, for example a thermosetting polymer may have the best properties for a product but it is not recyclable. Therefore, the designer may choose a thermoforming polymer with properties as close to the thermosetting one as possible.

## Justification of the choice

- The designer must make informed choices based on these characteristics and other factors, such as:
  - cost – is there another material available that may cost less?
  - availability – is the material in good supply?
  - sustainability – is the material sustainable?
  - environmental impact – can the material be recycled?
  - social, cultural and moral considerations – will the use of this material have a negative impact?
- When designing an electronic system, the designer must consider the range of components available and decide on the most suitable type for the application.

---

### Now test yourself

TESTED

1  State **four** important mechanical properties of materials.  (4)
2  State **two** factors that can influence and affect the properties of materials.  (2)

---

# 1.14 Environmental, social and economic challenges

## 1.14.1 Respect for different social, ethnic and economic groups

REVISED

### Positive and negative impacts of products

- Products can have a negative impact on society that the designer may not have foreseen.
- For example, leaflets and flyers can be used to communicate useful information to people about services in the area, however this can lead to unwanted junk mail being delivered, urging people to buy things they don't want.
- While it is not always easy to predict the effects of a product, designers must try to consider the possible negative implications and decide whether the product is worthwhile.

## Cultural awareness

- **Cultural awareness** is considering how different cultural groups may be affected by a product. Different cultures may view products in a totally different light depending on their beliefs, ideas and experience.
- A product designed for a particular market sector should not be offensive or upsetting to another.
- Designing new products that continue traditional processes, skills and materials can help preserve cultural identity.

> **Cultural awareness**: considering how different cultural groups may be affected by a product.

## Economic challenges

- The manufacturing processes, transportation, use and disposal of a product can affect the economic needs of people.
- The manufacture of a new product may create jobs in a factory or geographical area, boosting local people's income and improving the overall economy of the area.
- If the product is manufactured with modern CAM machines, it could lead to the loss of jobs and, in turn, affect the whole economy in that area.
- Many products are made abroad because labour costs are much cheaper in developing countries.
- Some workers in these countries are poorly paid, work in dangerous or unhealthy conditions, and work long hours without breaks or holidays.
- In some countries there is use of child labour, with children working in unsafe conditions because they have no choice.
- Designers have a responsibility to ensure that people are not being exploited in this way.

# 1.14.2 Appreciation of the environmental, social and economic issues

REVISED

- Designers should make sure their clients follow the guidelines for the fair treatment of workers set out by organisations such as the European Institute for Crime Prevention and Control, the Gangmasters and Labour Abuse Authority (GLAA) or the Alliance for Workers Against Repression Everywhere (AWARE).
- Manufacturers of fairtrade products ensure that their workers are treated and paid fairly, and that they do not work in dangerous or unhealthy conditions (see Section 1.2.3 Ethical perspectives on page 12).
- Carbon offsetting is when companies that use high amounts of energy or produce a lot of waste 'offset' this by setting up or donating funding to projects that help to reduce greenhouse gases.
- Products that can be disassembled easily allow the components to be reused instead of throwing away perfectly useable parts and manufacturing new ones. Making it easier to separate parts for recycling means that less materials are disposed of in landfill.

### 1.14.3 Green design

- Social pressure to have the latest products and the influence of fashion and trends mean that people throw away rather than repair things. This is known as the **throwaway society**.
- **Green design** means using eco-friendly materials and construction practices that safeguard air, water and earth.
- **Eco-design** means designing sustainable products that will not harm the environment by considering the effects of the technology, processes and materials used.

### 1.14.4 Recycling and reusing materials or products

REVISED

- Choose materials that are sustainable, recyclable and non-toxic, and which do not require as much energy to process.
- Design products to last as long as possible, so there is less replacement of parts, and that can be fully recycled at the end of their useful life.
- Design products that function to the best of their potential.
- Consider the needs and wants of all the **stakeholders** involved.

### 1.14.5 Human capability

REVISED

- Human capability is the range of functions that humans can carry out.
- Cognitive capability is the ability to process thoughts, learn new things and solve problems.
- Physical capability is the ability to perform physical tasks, such as walking, standing, balancing or gripping items.
- As we mature from children into adults, our cognitive and physical capabilities develop and increase.
- Old age, illness or other events in our life can have a huge effect on our human capability.

### 1.14.6 Cost of materials

REVISED

- Designers must try to keep material costs as low as possible without compromising the quality of the final product.
- Where materials are expensive, designers must look for cheaper alternatives with similar properties.
- Designers must consider whether a particular material is necessary or just for 'show'.

### 1.14.7 Manufacturing capability

REVISED

- Manufacturing capability is what a company or manufacturing plant is able to produce.
- Many companies have approved ways of producing a particular product.
- Manufacturing capability relates to the:
  - technological capability of the manufacturer – how up to date the equipment is

**Throwaway society**: buying products for brief use then disregarding them or throwing them away with little thought of the environmental impact.

**Green design**: using eco-friendly materials and practices to design products.

**Eco-design**: designing sustainable products that have a minimal effect on the environment.

**Stakeholder**: a person other than the main user who comes into contact with or has an interest in the product.

**Typical mistake**

Sustainable products and materials are not the same as recyclable products and materials.

- physical size and weight of the product being manufactured
- production capacity of the manufacturing plant.

## 1.14.8 Life-cycle analysis

REVISED

- The life cycle of a product is the stages a product goes through from beginning to end.
- A life-cycle analysis (LCA) assesses the environmental impacts of a product at the following stages:
  - extraction of the raw materials
  - processing of the materials into useable form
  - manufacture of the product
  - distribution and transportation to retailers
  - use by the customer
  - repair and maintenance of the product
  - disposal or recycling of the product.

**Exam tip**

Practise drawing LCAs for a range of products.

### Now test yourself

TESTED

1 State **three** factors that may be considered when designing recyclable products. (3)
2 Name **three** sustainable materials and explain why they are sustainable. (6)
3 Describe **one** factor that can influence human capability. (2)

# 1.15 The work of past and present professionals and companies

## 1.15.1 Analysing a product to specification criteria

REVISED

**Table 1.13** Specification criteria

| Criteria | Explanation |
|---|---|
| Form | The shape and appearance of a product |
| Function | The purpose of a product; what it will be used for |
| Client/user requirements | The needs, wants and values that are important to the people who will use it |
| Performance requirements | Factors that will help it to do its job; how it will perform when in use |
| Materials and components/systems | What the product is made from and why; the properties the materials have that support function; how the product operates (system) |
| Scale of production and cost | The number of products to be manufactured will affect the scale of production and costs |
| Sustainability | The environmental impact from origin to disposal |
| Aesthetics | The visual appearance of the product, for example colour, texture |
| Marketability | How likely the product will sell; features that will help product sales |
| Innovation | Does it have any radically new features or include the latest technology? |

## 1.15.2 The work of past and present designers and companies

REVISED

### Alessi

- Alessi has a reputation for designing and manufacturing desirable, high-quality everyday items such as kitchenware and tools.
- Its products are designed to be different, unusual, timeless and fun pieces.
- Originally Alessi manufactured products in stainless steel and other metals but, over time, materials such as porcelain, glass, wood and polymers have been used.
- Alessi has worked **collaboratively** with external designers such as Philippe Starck who designed the futuristic Juicy Salif fruit squeezer, and Michael Graves who designed the **iconic** kettle with a bird-shaped whistle.

### Apple

- Much of Apple's success is attributed to the importance they place on innovation and design.
- Apple were pioneers in graphical user interfaces (GUIs). The 1983 Apple Lisa computer was the first to have icons that represented files and folders, and a cursor controlled by a mouse.
- British industrial designer Jonathan Ive was responsible for styling the first iPod, iMac, iPhone and iPad.
- Apple products are easily distinguishable by their sleek design, consistent shapes, colours and materials. Aesthetics and user experience are at the forefront of Apple's design philosophy.
- Apple is often criticised for products being developed with **planned obsolescence**, for software updates that do not work on older products, and for developing their own ports for connecting other devices.

### Heatherwick Studio

- Thomas Heatherwick founded Heatherwick Studio in 1994. His background is in designing and making, and he is known for his innovative use of materials and engineering.
- Heatherwick Studio employs over 250 people from a wide range of design disciplines including architects, engineers, landscape architects and product designers who work collaboratively.
- The company works on projects that will have the greatest social impact. Their ethos is to create interesting and unusual spaces that are seen to be different.
- Each project is viewed as a separate problem. The multi-skilled workforce work together, everyone contributes ideas and a number of potential solutions will be considered.
- The client plays a critical role in the design journey.
- Notable projects include: Guy's Hospital, London; 1000 Trees, Shanghai; Pacific Place, Hong Kong; UK Pavilion, Shanghai; The Vessel, New York.

**Collaboration**: working with others on a task.

**Iconic**: a noteworthy design, style or product that is highly original or unique.

**Figure 1.30** Juicy Salif fruit squeezer by Philippe Starck

**Planned/built-in obsolescence**: when a product is designed to no longer function or be less fashionable after a certain period of time.

**Exam tip**

In an examination be able to describe the key features of a designer's work, show knowledge of their style and influences, and apply it to the products they designed.

**Figure 1.31** UK Pavilion, Shanghai

## Joe Casely-Hayford

- Joe Casely-Hayford (1956–2019) launched his first collection in 1983 under the label KIT. It was made from WWII army tents that were **disassembled**, made into clothing and sold to specialist fashion shops.
- The style reflected a worn and distressed look that proved both popular and successful. This method of making clothes was labour intensive, however.
- His next venture was to make shirts from robust cotton sourced from a mill in Manchester. These heavy-duty shirts, which opened at the front and back, were again successful.
- His work reflected his Saville Row tailoring experience and his free spirit as a designer. He embraced different cultures and his Ghanaian heritage through his work, merging innovative techniques and aesthetic styling.
- Many in the fashion world saw him as a talented pioneer – the first black designer to represent London on the international fashion stage.

**Disassembled**: taken apart piece by piece.

## Pixar

- Pixar began as the Graphics Group of the Computer Division of Lucasfilm in 1979. Its aim was to develop state-of-the-art technology for the film industry.
- It was bought by Apple co-founder Steve Jobs in 1986 and was established as an independent company called Pixar. Walt Disney bought Pixar in 2006.
- Pixar and Disney collaborated on a computer animation production system (CAPS) that revolutionised the way animated films were made.
- *Luxo Jr.*, a short film based on one large and one small desk lamp, was the result of this partnership and it received an Oscar nomination for best short (animated) film.
- Pixar is known for CGI-animated, feature-length films created with RenderMan – a photorealistic 3D rendering software that continues to be developed to meet the challenges of 3D visual effects.
- Pixar's first computer-animated feature film was *Toy Story* (1995). Other notable films include *A Bug's Life*, *Finding Nemo* and *The Incredibles*.

## Raymond Loewy

- Raymond Loewy (1893–1986) was one of the most influential industrial designers and is credited with revolutionising the industry, working on products ranging from postage stamps to cars and spacecraft.
- Much of his success was down to his philosophy of streamlining designs and applying functional styling.
- His work reflected his own design principle: MAYA – Most Advanced Yet Acceptable. He believed that if a design change was too radical it might not be acceptable to the consumer, but that should not prevent the development of a product to its full potential.
- His car designs are considered classics. He was an advocate for fuel efficiency long before the concerns we share today.
- Loewy was a highly talented and successful commercial artist. One of his most noted logos is for the Shell Oil Company. He believed a logo should be easily recognised and remembered, even if viewed only momentarily.
- Other notable works include the Coca-Cola bottle, John F. Kennedy's Air Force One and the American Greyhound bus.

## Tesla

- Tesla, an American electric automotive and energy company, is developing affordable, mass-market electric cars and battery products.
- Co-founder Elon Musk overseas all product development, from engineering to design.
- Tesla's first electric car, the Roadster, was introduced in 2008 and featured cutting-edge battery technology.
- The Model S, launched in 2012, combined safety, performance and efficiency, with the longest range of any electric vehicle (about 300 miles).
- SolarCity, a subsidiary of Tesla, manufactures and installs solar roofs and batteries, allowing homeowners and businesses to manage renewable energy generation and storage.

## Zaha Hadid

- Zaha Hadid (1950–2016) studied mathematics at the American University of Beirut before moving to the Architectural Association School of Architecture in London in 1972.
- Hadid opened her own architectural firm in London in 1980. Her modern designs stood out. Much of her early work was quite radical and, although acclaimed and published in many journals, was never built.
- In 1993 Hadid designed a small fire station for the factory of the Swiss furniture maker Vitra. The design, comprised of raw concrete and glass with sharp diagonal forms colliding at a focal point, launched her career as an architect.
- Hadid's style did not reflect any other architectural movements. She was referred to as the 'Queen of the Curve' but was also known for her use of geometric shapes. Much of her work was futuristic, combining curved edges with sharp angles in concrete and steel, using the latest materials and technological advancements.
- Notable work includes: the Heydar Aliyev Centre, Baku, Azerbaijan; Glasgow Museum of Transport, Scotland; and the National Museum of the 21st Century Arts (MAXXI), Rome.

**Figure 1.32** The Heydar Aliyev Centre, Baku, reflects Hadid's distinctive architectural style

**Typical mistake**

Read questions carefully. If a question is about a designer or a company, discuss their work. Biographical facts about a designer or company are not needed and will not gain credit.

### Now test yourself     TESTED ☐

1  Describe **two** characteristics of Alessi products.     (2)
2  Joe Casely-Hayford trained in Saville Row. Explain how this influenced his style of work.     (2)
3  State what was significant about the Tesla Roadster that was launched in 2008.     (1)
4  Describe what was significant about the development of RenderMan technology.     (2)
5  Explain why Zaha Hadid's approach to design differed from conventional architects' style.     (3)

# 1.16 Design strategies

## 1.16.1 Use of different design strategies

REVISED

- Innovative design is about exploring new ways of doing things or trying to approach problems from different angles or perspectives.
- **Design fixation** is when designers stick to one idea instead of trying to explore new design avenues. Designers can use different strategies to help them avoid design fixation.

### Collaboration

- A design business will usually employ a number of designers. By discussing, sharing and working with each other they can get a different perspective on a problem.
- **Collaboration** allows designers to bounce ideas back and forth between each other. It can open new avenues of investigation and designs that one designer would not have thought of on their own.

### User-centred design

- **User-centred design** is where the needs, wants and requirements of the **user** are looked at and checked at every stage.
- Putting the user at the centre of the design process means less refinement and modifications are likely to be needed later.
- The process has four main stages:
  1 **Context of use**: identify the users of the product, what they will use it for, and under what conditions it will be used.
  2 **Requirements**: identify any user goals that must be met.
  3 **Create design solutions**: this may be done in stages, building from a rough concept to a complete design.
  4 **Evaluate designs**: evaluation through usability testing with actual users.

### Systems thinking

Systems thinking is a top-down method of designing, analysing or describing how the parts of a system work together to achieve a particular function. These parts are often broken down into input, process and output subsystems. In design and technology, this strategy is applied mainly to electronic and mechanical systems but can be used to help understand other, wider human-made and natural settings, for example natural ecosystems.

> **Design fixation**: when designers stick to one idea instead of exploring new design avenues.
>
> **Collaboration**: working with others on a task.

> **Exam tip**
>
> Practise using different design strategies in class. This will help you understand the benefits of each.

> **User-centred design**: the needs and wants of the user are considered at every stage.
>
> **User**: the person or group of people a product is designed for.

---

### Now test yourself

TESTED

1 Explain **one** benefit of collaborative design work. (2)
2 Describe how user-centred design can save time when designing products. (2)

# 1.17 Developing, communicating, recording and justifying design ideas

## 1.17.1 Using a range of communication techniques and media

### Sketching

- **Freehand sketching** is a method of presenting design ideas on paper without the use of drawing aids or formal drawing conventions. It is often used to produce early ideas for products or as a tool for ideas generation. Sketching can be done in either 2D or 3D.
- Annotations are used to give further details and information about the ideas that have been sketched. This could cover areas such as:
  - the inspiration behind each design
  - the good and bad points
  - manufacturing methods that could be used to make each idea
  - how well each design meets the requirements of the **end user** and/ or design specification.

### Cut and paste techniques

Cut and paste techniques are used to make collages. Pictures and images are cut from different sources, such as newspapers, magazines and photographs, and glued on to paper to create an art form.

### Digital photography/media

Digital technologies are increasingly being used in the design of products. For example, taking photos of iterative models to record each stage of development, or videos of testing to demonstrate how a design works. These can be shared online with the end user or other potential customers.

### 3D models

The use of **3D models** allows design ideas to visualised, handled and tested. Models allow the designer to spot any errors or areas for improvement before money is spent on more expensive materials. They can also be shared with the end user to gain their feedback before the final product is manufactured.

### Isometric and oblique projections

- These are methods of drawing objects in 3D.
- In **isometric projections** the sides are drawn at a 30-degree angle.
- Oblique projections are simpler to produce than isometric, but not as realistic. The front side of the object is drawn first and lines are then drawn backwards at 45 degrees. The object is inclined either to the left or to the right.

> **Freehand sketching**: method of presenting design ideas on paper without the use of drawing aids or formal drawing conventions.
>
> **End user**: the person who will purchase or use the finished product.

**Figure 1.33** 3D sketching

> **3D model**: 3D version of a design idea made using cheap materials, for example card and Styrofoam™, used to check how it would look and function.
>
> **Isometric projection**: method of drawing objects in 3D by drawing the sides at a 30-degree angle.

# Perspective drawing

- **Perspective drawings** show how objects get smaller as they are further away, thus in 'perspective'.
- In one-point perspective, the objects being drawn converge towards a single vanishing point on a horizon line. In two-point perspective, there are two vanishing points on either side of the horizon line. This produces more realistic drawings.

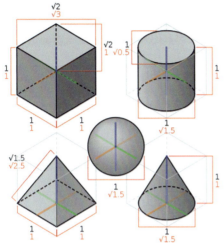

**Figure 1.34** A number of basic forms in isometric projection

**Figure 1.35** A railway station drawn using one-point perspective

## Orthographic and exploded views

- **Orthographic projections** are working drawings that show 3D objects in 2D. This is achieved through showing the front, plan and side views of the product.
- In first angle orthographic projections, the front view is drawn above the plan view, and the side view is drawn to its left.
- In third angle orthographic projections, the front view is drawn below the plan view, with the side view to its right.
- There are standard lines that must be used for different purposes when producing orthographic drawings.
- An exploded view shows the parts of a product suspended in space and slightly separated by distance. It shows the relationship between all the various parts of the product.

**Typical mistake**

Be careful not to confuse first and third angle orthographic projections.

**Orthographic projection**: method of showing a 3D object in 2D, using front, pan and side views.

**Figure 1.36** An exploded view of a sweet dispenser

## Assembly drawings

**Assembly drawings** show how all the different parts of a product fit together. Each part is identified by a number. A bill of materials is included that lists each part needed and the quantities required.

## System and schematic diagrams

- System, or block, diagrams show a system in terms of its input, processes and output subsystems. This presents a top-down overview of the system.
- **Schematic diagrams** show a circuit in terms of the individual components used. These are represented using symbols; the connections between them are shown with straight lines.

## Computer-aided design

CAD software can be used to produce both drawings and virtual models of designs. Using CAD has many advantages, such as:

- Accuracy of drawing and modelling is very high.
- Changes can be made quickly and easily.
- No materials or physical resources are needed, reducing costs.
- Models can be simulated on screen.
- 2D drawings can be produced directly from the 3D model, saving time.
- CAD files can be outputted directly to CAM equipment for manufacture.

> **Assembly drawing**: drawing that shows how all the different parts of a product fit together, with each part identified by a number.
>
> **Schematic diagram**: uses symbols to show a circuit in terms of its individual components and how they are connected.

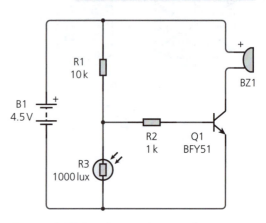

**Figure 1.37 An electronic transistor circuit schematic**

> **Exam tip**
>
> Make sure you know the purpose and correct use of each of the various ways of developing and communicating design ideas.

## 1.17.2 Recording and justifying design ideas clearly and effectively

REVISED

- Designers must be able to justify their ideas and the design decisions that they have taken, for example giving reasons why certain materials and components have been chosen. This should always be in terms of the needs and wants of the end user and how well each design would meet them.
- This can be achieved through detailed annotations of design ideas, or listing how each one meets the different requirements of the design specification. Consideration should be given to the wider design context and how each idea fits within this.

> **Now test yourself**
>
> TESTED
>
> 1 Describe the differences between freehand sketching and formal drawing. (4)
> 2 Name **two** types of perspective drawings. (2)
> 3 Explain the differences between first and third angle orthographic projections. (4)
> 4 Describe the purpose of a systems diagram. (2)
> 5 Give **two** reasons why CAD is used to produce drawings of products. (2)

## Exam practice

1 Sustainability and environmental issues are important considerations when developing new products.
   a Define the term 'sustainable design'. (2)
   b Explain how the circular economy benefits the environment. (4)
   c Describe how studying a product's life cycle benefits a manufacturer. (3)
2 Smart materials can change their properties or appearance in response to external stimuli.
   a State what makes shape memory alloys (SMAs) different to other metals. (1)
   b Complete the following sentence: Nitinol is one of the most common shape memory alloys; it is made from .............. and .............. (2)
   c Describe an example where thermochromic pigment used in a named product would benefit the user. (3)
3 Electronic devices can be embedded into textile fabrics for clothing.
   a Heart rate monitors can be embedded into fabric. Describe how this would benefit a user with a heart condition. (2)
   b Micro-encapsulation is often used in medical textiles. Give **two** examples of its use in medicine and explain the benefit to the patient. (6)
   c Explain why Rhovyl® is used in sportswear. (2)
4 Fibres are the raw materials of textiles.
   a State one source of natural textile fibres and one source of synthetic textile fibres. (2)
   b Cotton is widely used in the manufacture of textile products. Discuss the reasons why it is used in clothing. (4)
5 An outdoor street lamp must light up automatically when it gets dark. Name a suitable sensor and output device for this system. For **each** component named, explain how it functions to achieve the required outcome. (4)
6 A V-belt system has a driver pulley with a diameter of 300 mm and a driven pulley with a diameter of 100 mm. Calculate the velocity ratio of the system. (3)
7 Explain the purpose of a systems diagram. (2)
8 Explain why mild steel can be considered to be an alloy as well as a ferrous metal. (3)
9 What material is used to reinforce the polyester resin in glass reinforced plastic (GRP) and how does it improve the property of the resin? (2)
10 What is the difference between 'toughness' and 'hardness' when referring to timbers? (4)

# 2 Metals

## 2.1 Design contexts

Metals provide the designer with a group of materials that have an extensive range of properties. It is important to have an understanding of metals so that you can make an informed judgement as to when and where they should be used.

## 2.2 Sources, origins, physical and working properties, and social and ecological footprint

Metals are categorised as ferrous or non-ferrous – **ferrous metals** contain iron and **non-ferrous metals** do not.

### 2.2.1 Ferrous metals (including 2.2.3 Sources and origins, 2.2.4 Physical characteristics and 2.2.5 Working properties)

REVISED

> **Ferrous metals**: metals that contain iron.
>
> **Non-ferrous metals**: metals that do not contain iron.
>
> **Tensile strength**: the ability to withstand being pulled apart.
>
> **Compressive strength**: the ability to withstand a pushing (squashing) force.

Table 2.1 Common ferrous metals

| Metal | Geographical source | Physical and working properties | Composition | Melting point | Uses |
|---|---|---|---|---|---|
| Mild steel | China | Good **tensile strength**, malleable, tough, magnetic, easily joined, poor corrosion resistance | Iron, carbon (0.1–0.3%) | 1400 °C | Nuts, bolts, washers, screws, car bodies, rolled steel joists (RSJs) |
| Stainless steel | China | Excellent tensile strength, very tough, very durable, excellent resistance to corrosion | Iron, carbon, chromium (18%), nickel (8%) | 1400 °C | Kitchen sinks, cutlery, surgical equipment |
| Cast iron | USA, Russia, Sweden | Hard, brittle, good **compressive strength**, magnetic | Iron, carbon (3.5%) | 1200 °C | Metalwork vices, brake discs, drain and manhole covers |

→

Answers, glossary and quick quizzes at www.hoddereducation.co.uk/myrevisionnotes

**Table 2.1** Common ferrous metals (continued)

| Metal | Geographical source | Physical and working properties | Composition | Melting point | Uses |
|---|---|---|---|---|---|
| High-carbon steel | China | Very high strength, very hard, magnetic, can be heat treated | Iron, carbon (0.8–1.5%) | 1800°C | Cutting tools, such as saw blades and drill bits |
| Tungsten steel | China | Excellent strength, excellent resistance to corrosion, very tough | Iron, tungsten (1–18%) | 1650–3695°C | High-speed cutting tools, turbine blades |

## 2.2.2 Non-ferrous metals (including 2.2.3 Sources and origins, 2.2.4 Physical characteristics and 2.2.5 Working properties)

REVISED

**Table 2.2** Common non-ferrous metals

| Metal | Geographical source | Physical and working properties | Composition | Melting point | Common uses |
|---|---|---|---|---|---|
| Aluminium | USA, France, Australia | Lightweight, soft, ductile, malleable, good conductor of heat and electricity, corrosion resistant, excellent strength-to-weight ratio | Pure metal | 660°C | Drinks cans, cooking pans |
| Copper | USA, Chile, Zambia, Russia | Extremely ductile and malleable, excellent conductor of heat and electricity, easily soldered, corrosion resistant | Pure metal | 1084°C | Plumbing fittings, hot water tanks, electrical wire |
| Brass | USA, Chile, Zambia, Russia | High resistance to corrosion, excellent conductor of heat and electricity, excellent casting properties, harder than copper | Copper (65%), zinc (35%) | 900–940°C | Plumbing fittings, door fittings, musical instruments |
| Tin | Indonesia, China | Soft, ductile, malleable, high resistance to corrosion, weak | Pure metal | 232°C | Metal coatings, solder |
| 7000 series aluminium | USA, France, Australia | Excellent strength-to-weight ratio | Aluminium, zinc, magnesium, copper | 660°C | Aircraft frames, alloy wheels, bike frames |
| Titanium | China, Japan, Russia, USA | High resistance to corrosion, excellent strength, tough, low density | Pure metal | 1668°C | Dental implants, replacement human joints |

## 2.2.6 Social footprint

REVISED

### Trend forecasting

- Trend forecasting is concerned with predicting the need for metal products in future years.
- Trend forecasters carry out research by surveying target markets, interviewing focus groups and investigating new developments.

### Impact of extraction and production on communities and wildlife

- Most metals are embedded within ore, which is found in the Earth's crust. Ore has to be mined to extract it from the ground.
- Mining ore provides employment for local people and can lead to the development of new communities. New roads and infrastructure are produced to service the communities.
- **Deforestation** can occur to make way for the mine. Wildlife can be disturbed and endangered species lost if the mining process is not carried out in a sensitive manner.
- Mining can also lead to the pollution of rivers and water courses.

**Deforestation**: the large-scale felling (cutting down) of trees that are not replanted.

### Recycling and disposal

- Most metals can be recycled. Around 9.5 billion aluminium drinks cans are produced in the UK each year, 75 per cent of which are made from recycled aluminium.
- Some metals are difficult to recycle, such as the copper within electrical cables.
- Some metals, like lead and mercury, are toxic. If they are allowed to go to landfill, they will pollute the land.
- Most metals are non-biodegradable and will stay in landfill forever.

**Typical mistake**

When asked to describe the life cycle of a metal product, candidates often start with the actual product and fail to give details of the sourcing and processing of the raw material.

## 2.2.7 Ecological footprint

REVISED

### Sustainability

- Metal is a finite resource – once we have used it all up, there is no more.
- Most metals are recyclable; if we use them carefully, they should never run out.

### Extraction and erosion of the landscape

- Open cast mining causes scars on the landscape.
- Deforestation can lead to soil erosion.

**Figure 2.1 Open cast mining**

## Processing

- Iron ore needs to be smelted in a blast furnace to release the metal.
- Smelting involves the burning of fossil fuels to heat the ore to a very high temperature.
- Burning fossil fuels releases carbon dioxide into the atmosphere, which is a major contributor to global warming and climate change.

## Transportation

- Ore needs be transported to the blast furnace by road, rail and/or by sea.
- Each method involves the burning of fuel, which gives off carbon dioxide.

## Wastage

- Most waste metal can be recycled.
- Waste products from blast furnaces can be crushed and used as filler in cement.

## Pollution

- The extraction, processing and transportation of metal creates a significant amount of pollution, mainly in the form of toxic gases such as carbon dioxide being released into the atmosphere.
- Metal that is allowed to go to landfill can pollute the land and contaminate water courses.

> ### Now test yourself   TESTED ☐
>
> 1  Give **two** important properties of stainless steel.   (2)
> 2  What is the difference between a ferrous metal and a non-ferrous metal?   (1)
> 3  Give **two** important properties of aluminium.   (2)
> 4  Explain the ecological impact of mining metal ore.   (3)
> 5  Discuss the sustainability of using metal.   (3)

# 2.3 Selection of metals

## 2.3.1 Aesthetic factors
REVISED ☐

- The form of a product relates to its shape, size and proportion. Metal can be shaped, moulded and cast into an infinite number of forms.
- Metal is available in several different natural colours. Steel has a grey appearance while copper and brass are yellowy-orange. Metals can also be coated in different ways to change their appearance.
- The **texture** of metal is generally smooth, but this can be altered during manufacture. You may have noticed that the floor of the bus you came to school on was covered in texture metal plate. This gives it grip so that you don't slip.

**Figure 2.2 Checker plate**

> **Aesthetics**: how something looks and feels.
>
> **Texture**: the feel and finish of a material.

### 2.3.2 Environmental factors

- Metals can be considered to be a sustainable material if they are recycled at the end of their useable life. Their lifespan can be lengthened if thought is given to maintenance and repair during the design process.
- The extraction and processing of metals can lead to pollution as carbon dioxide and other toxic chemicals are released into the atmosphere. Metals that go into landfill can pollute the land.
- The processing of metals requires a lot of energy; for example, fossil fuels are burnt to heat iron ore to very high temperatures in order to produce steel.
- Unprotected steel will corrode if not protected. The corrosion appears as rust and can lead to the failure of a steel structure.

### 2.3.3 Availability factors

- Metal can be purchased in a variety of standard **stock forms**. You can save time cutting and shaping metal if you design around the available sizes and shapes.
- Many specialised metal alloys have unique properties; for example, adding chromium to steel will prevent it from rusting.
- The price of metals can vary according to their rarity and market demand. Steel is relatively inexpensive as it is produced in vast quantities; gold is expensive as it is very rare.

> **Stock forms**: the standard shapes and sizes that a material or component is available in.

### 2.3.4 Cost factors

- The quality of a material will affect its cost. High grades of aluminium used within the aircraft industry require greater processing and therefore come at a higher cost.
- The scale and method of production will significantly affect costs. A wrought iron gate handcrafted by a highly skilled blacksmith will be very expensive; a steel food can, produced by machine, will be relatively inexpensive.
- World metal prices are referenced on the London Metal Exchange. These prices reflect the true cost of metals, taking into account supply, demand, currency values, global tension and availability.
- Recycling steel has a significant effect on the cost of a steel product. By recycling you cut out the considerable costs of sourcing, mining, processing and transporting iron ore.

### 2.3.5 Social factors

- Different social groups make different demands on metal products. Mountain bikers have seen a change from steel-framed bikes to aluminium-framed bikes, and now there is a growing social demand for carbon fibre-framed bikes.

- Changing trends and fashions can influence the choice of metals. The metal used to manufacture kitchen pans can vary from stainless steel, aluminium or cast iron depending on the style, trend and fashion at the time of purchase.
- The popularity of a metal product can influence its cost, as it is generally more economical to manufacture in high quantities.

## 2.3.6 Cultural and ethical factors

REVISED

- It is important that products do not offend any particular culture. Offensive products are also unlikely to sell.
- Metal-based products should be suitable for the intended user. For example, a child's toy made from cast iron would be too heavy and potentially dangerous.
- A consumer society, where products are constantly changing to keep up with fashions and trends, can lead to a high proportion of unwanted products. These products should be sold or passed on to other users, or recycled.
- The increase in mass production allows for a greater number of affordable metal products to be produced. There are approximately 500,000 more cars on the road each year in the UK.
- Many companies now utilise **planned obsolescence** to ensure that there will be an ongoing demand for their products. A metal component may be designed to break after a certain number of uses, forcing the consumer to purchase a new product.

> **Exam tip**
>
> When a question asks you to **discuss** your answer, the examiner needs you to look at both sides of an argument.

> **Planned/built-in obsolescence**: when a product is designed to no longer function or be less fashionable after a certain period of time.

> **Typical mistake**
>
> When asked to evaluate or discuss, candidates will typically provide a conclusion but fail to justify. Make sure you provide detailed reasoning in your answer.

## Now test yourself

TESTED

1 What is the colour of brass? (1)
2 What happens to steel if it is left unprotected? (1)
3 If a metal is in high demand, what will happen to its price? (1)
4 How can ethical factors affect the design of metal products? (2)
5 What is 'planned obsolescence'? (2)

# 2.4 Forces and stresses

Metals are known for their strength and durability, but these properties can be further enhanced.

## 2.4.1 Forces and stresses

**Table 2.3 Forces and stresses acting on metals**

| Type of force | Definition | Example |
|---|---|---|
| Compression | A force that is pushing down on an object | A chair or table leg |
| Tension | A force that is pulling an object apart | Cables on a suspension bridge |
| Shear | A force that acts in opposite directions | Scissors cutting paper |

**Compression force**: a pushing force.

**Tension force**: a pulling or stretching force.

**Shear force**: a force acting in opposite directions.

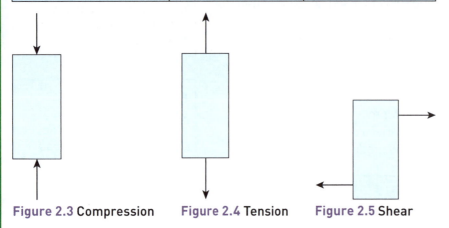

**Figure 2.3 Compression**     **Figure 2.4 Tension**     **Figure 2.5 Shear**

- An electrical current can be used to produce a rotary force when it is passed through an electrical motor.
- Ferrous metals can be attracted or repulsed by a magnetic force.

## 2.4.2 Reinforcing and stiffening techniques

### Hardening

Ferrous metals with a medium- to high-carbon content can be made even harder by a heat treatment process known as **hardening**:

1 Heat the steel with a brazing torch until it reaches red heat.
2 Quickly take the red-hot steel and quench it in cold water.

**Hardening**: the process of heat-treating metal to make it hard.

### Tempering

- Hardening, as the name suggests, hardens steel but also makes it brittle. If it is left in this state, the steel component could snap or shatter in use.
- To reduce the brittleness, but retain most of the hardness, the steel should undergo a process known as **tempering**:
    1 Clean the steel with emery cloth until it is bright silver.
    2 Heat the steel with a brazing torch until it reaches the correct tempering colour (see Figure 2.6).
    3 Quickly take the heated steel and quench it in cold water.

**Tempering**: the process of heat-treating metal to reduce its brittleness.

| Colour | Temp. (°C) | Hardness | Typical uses |
|---|---|---|---|
| Light straw | 230 | Hardest | Lathe tools, scrapers |
| Dark straw | 245 | | Drills, taps and dies, punches |
| Orange/brown | 260 | | Hammer heads, plane irons |
| Light purple | 270 | | Scissors, knives |
| Dark purple | 280 | | Saws, chisels, axes |
| Blue | 300 | Toughest | Springs, spinners, vice jaws |

**Figure 2.6** The tempering colours of steel

**Typical mistake**

Candidates often fail to use the correct technological terminology when answering questions on hardening and tempering. Make sure you understand these terms.

## Effect of carbon content

Increasing the carbon content of steel makes it significantly harder.

**Table 2.4** The effect of carbon content

| Steel type | Carbon content | Uses |
|---|---|---|
| Mild steel | 0.15–0.35% | Car bodies, building framework |
| Medium-carbon steel | 0.4–0.75 | Screwdrivers, pliers, hammers |
| High-carbon steel | 0.8–1.5% | Saws, drills, chisels |

## Work hardening

- Work hardening occurs when metals are bent, rolled or hammered into shape.
- The metal becomes hard and brittle, and will eventually crack.
- The metal can be **annealed** to remove the effects of work hardening:
  1. Heat the steel with a brazing torch until it reaches red heat.
  2. Allow the metal to cool down slowly.

**Exam tip**

When asked to **describe** a process, make sure you use both notes and labelled sketches in your answer.

## Steel construction beams

Steel construction beams are used extensively in the building industry. The cross-sectional shape of a steel construction beam provides strength in different directions.

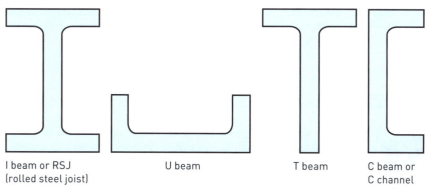

I beam or RSJ
(rolled steel joist)     U beam     T beam     C beam or
C channel

**Figure 2.7** Constructional steel beams

**Now test yourself**

TESTED

1  What is meant by 'tempering'? (1)
2  What is meant by a 'shear force'? (1)
3  Where and how would you use an RSJ? (2)
4  What is the effect of increasing the carbon content of steel? (2)
5  Describe the process of hardening steel. (3)

# 2.5 Stock forms, types and sizes

## 2.5.1 Stock forms/types

REVISED

**Stock form** is the name given to the standard form that metal can be purchased in from suppliers. It is cost effective to produce designs that incorporate stock forms.

> **Stock forms**: the standard shapes and sizes that a material or component is available in.
>
> **Plate**: thick sheets of metal.

**Table 2.5** Stock forms of metals

| Stock form | Image | Description | Use |
|---|---|---|---|
| Bar | | A metal bar with a solid cross-section. Usually sold in 4 m lengths | General purpose fabrication work |
| Sheet | | 2440 mm × 1440 mm sheet of metal. Usually available in varying thickness up to 3 mm | Car bodies. Casings for 'white goods' (washing machines, fridges, dishwashers) |
| **Plate** | | 2440 mm × 1220 mm metal plate. Usually available in varying thicknesses over 3 mm | Ships hulls. Anchor points |
| Pipe and tube | | Metal pipe and tubing with a hollow cross-section | Water pipes, structural framework |
| Casting | | Complex shapes formed by heating ingots of metal and pouring the molten metal into a mould | Metalwork vices and car engines |

→

**Table 2.5** Stock forms of metals (continued)

| Stock form | Image | Description | Use |
|---|---|---|---|
| **Extrusion** | | Complex cross-sectional shapes formed by pressing a **billet** of metal through a die | Curtain rail |
| Wire | | A billet of metal is drawn through a die to produce very thin, circular cross-sections | Electrical cable |
| Powdered metallurgy (sintering) | | Powdered metal is heated and pressed into a mould | Bearings and gears |

> **Extrusions**: long lengths of metal with a consistent cross-section.
>
> **Billet**: a stock form of metal used when casting or extruding.

## 2.5.2 Sizes

REVISED

Metal can be measured in several different ways.

**Table 2.6** Sizes

| Size | Diagram | Description |
|---|---|---|
| **Gauge** | 16 swg | The thickness of sheet metal is measured in a unit known as standard wire gauge (SWG)<br><br>16 swg = 1.626 mm |
| Cross-sectional area | 20<br>20 | The cross-sectional area is calculated by multiplying the width of the cross-section by the height of the cross-section<br><br>A circular cross-section is calculated by using the formula $\pi d$, where d = the diameter |
| Diameter | 20 | The diameter of round bar or tubing is the distance from one side of the circle to the other |

**Table 2.6** Sizes (continued)

| Size | Diagram | Description |
|------|---------|-------------|
| Wall thickness | | Wall thickness is calculated as the distance from the outside of the tubing to the inside |

**Gauge**: a unit of measurement for sheet metal.

### Now test yourself

TESTED

1  What is meant by the term 'stock form'? (2)
2  What is the name given to metal sheets over 3 mm in thickness? (1)
3  What is another name for powdered metallurgy? (1)
4  Draw a diagram to show the wall thickness of a tube. (2)
5  Draw a diagram to show the cross-sectional area of a square bar. (2)

**Typical mistake**

Candidates often lose marks because they have not used the correct units when answering a question. Check if you should give your answer in millimetres, centimetres or metres.

# 2.6 Manufacturing to different scales of production

## 2.6.1 Processes

REVISED

### Forging

- **Forging** steel involves heating the metal until it is bright red and then hammering it into shape.
- A blacksmith will produce one-off or small batch products, such as horseshoes and elaborate metal sculptures. Drop forging is an industrial process used to produce products such as spanners.
- **Advantages**: forging produces strong metal products with a tight, flowing grain structure.
- **Disadvantages**: forging is a specialist process; it can be time-consuming and is therefore relatively expensive.

### Casting

- Before you can **cast** metal, a mould must be made. When casting with pewter, the mould can be produced on a laser cutter from sheets of thin medium-density fibreboard (MDF). When casting with aluminium, the mould would be made in sand.

**Forging**: shaping metal by heating and hammering.

**Figure 2.8** A blacksmith at work

**Casting**: shaping by pouring molten metal into a mould.

- The metal is heated until it becomes molten and is then poured into the mould.
- The body of metalwork vices and car engines are produced by casting.
- **Advantages**: the mould can be used many times, producing identical products in a relatively short time.
- **Disadvantages**: the process is only cost effective when a high quantity of identical products are required.

## Powder metallurgy (sintering)

- Powdered metal is heated and compressed under immense pressure in a two-part mould.
- Bearings, gears and lathe cutting tools are produced by sintering.
- **Advantages**: identical products can be produced quickly by this method.
- **Disadvantages**: there is a high initial set-up cost as advanced equipment must be purchased.

## Stamping

- **Stamping** involves pressing thin metal sheet into shape.
- Car body panels are made by stamping.
- **Advantages**: identical shapes can be produced quickly.
- **Disadvantages**: the process is only cost effective when high volumes of identical products are required.

> **Stamping**: pressing a sheet of metal into a shape.

## Welding

- Welding produces a strong permanent joint between two or more metals.
- The metal components are heated locally until they become molten; the surfaces then fuse (melt) together.
- Oxyacetylene welding utilises a gas flame to heat the metal; MIG, TIG and arc welding use electricity to heat the metal.
- **Advantages**: welding produces a very strong, permanent joint.
- **Disadvantages**: welding requires a high level of skill.

**Figure 2.9 Welding a steel frame**

## Extrusion

- Extruding metal produces a long, continuous, intricate cross-section, such as curtain railing and tubing.
- A billet of metal is heated and forced through a metal die – think of squirting toothpaste out of a tube.
- **Advantages**: extrusion produces a long, consistent profile of metal.
- **Disadvantages**: the set-up costs are very high; it is only cost effective where high volumes of extruded metal are required.

## Hardening

See Section 2.4.2 Reinforcing and stiffening techniques, on page 56.

## 2.6.2 Scales of production

The scale of production is influenced by the demand for the product, its complexity and the intended purchase price.

**Table 2.7** Scales of production for metals

| Scale of production | Advantages | Disadvantages | Examples |
|---|---|---|---|
| One-off | A high-quality, unique product is produced | Usually involves a specialised process requiring a highly skilled workforce<br><br>Expensive | Wrought iron gates<br><br>Wedding rings |
| Batch | A limited number of identical products are produced<br><br>The cost of the product is reduced as materials can be bought in bulk and processing times are reduced | A high initial set-up cost involving the purchase of machinery and manufacturing aids | Handles<br><br>Hinges<br><br>Locks |
| Mass | Large quantities of identical products are produced<br><br>The cost is further reduced as computer-aided manufacture can be utilised | A high initial set-up cost as there is intensive use of machines<br><br>Products lose their individuality | Cars<br><br>Fridges<br><br>Microwaves |
| Continuous | The unit cost of a product is significantly reduced as materials can be purchased in vast quantities<br><br>A very high demand can be satisfied | Initial set-up cost is very high, often requiring a dedicated factory to be built; there is likely to be extensive use of robotics and computer-aided manufacture | Soft drinks cans<br><br>Production of steel |

## 2.6.3 Techniques for quantity production

### Marking out

- Metal is often coated with marking blue to help identify marking out lines.
- A scriber is used to scratch lines into the surface of the metal.
- A datum line is initially placed on the metal, from which all subsequent lines are measured.
- An engineer's square is used to produce a line 90 degrees to an edge.
- A centre punch makes a dent in metal prior to drilling.
- Odd leg callipers produce a line parallel to an edge.
- Dividers are used to draw a circle/arc on a metal surface.

### Jigs

- **Jigs** speed up production and increase accuracy by removing some of the marking out processes.
- A drilling jig can be placed on to a metal component while a series of holes are drilled.

**Jig**: a mechanical device to aid production.

## Fixtures

- A fixture speeds up production and increases accuracy by removing the need for setting up a process.
- A welding fixture will hold metal in the correct place while a frame is welded.

## Templates

**Templates** can be placed on a metal surface and drawn around, or they can be fixed to the surface of the metal and machined around. They reduce the need for marking out.

> **Template**: a 2D shape to aid cutting out a profile.

## Patterns

When casting in aluminium, a pattern is initially produced from a hardwood such as jelutong. The pattern is then used to produce identical sand moulds.

## Moulds

When casting in aluminium, molten metal is poured into a pre-prepared sand mould. When die casting, the mould would be made from steel.

## Sub-assemblies

Car manufacturers make extensive use of sub-assemblies. An engine will be put together as a sub-assembly before it is attached to a car.

## Computer-aided manufacture

Computer Numerically Controlled (CNC) machinery, such as lathes and milling machines, produce metal components accurately, consistently and cost efficiently.

> **Typical mistake**
>
> Candidates often confuse CAD (computer-aided design) with CAM (computer-aided manufacture). Make sure you know the difference.

## Quality control

Quality control checks should be carried out at regular intervals. This can involve removing products from a production line and checking them for dimensional accuracy.

> **Tolerance**: the allowable amount of variation of a stated measurement.

## Working within tolerances

- If a metal component is to interact with another component, it must be manufactured within a certain **tolerance** or it may be too big or too small to function. A tolerance is given a size plus or minus an acceptable limit.
- **Example**: if a hole was given a size of 100 mm ± 0.5 mm, the maximum size for the hole would be 100.5 mm and the minimum size would be 99.5 mm.

> **Exam tip**
>
> When producing a sequence of operations, make sure you have them in the correct order. If you notice later that you have left something out, add it to the bottom of your answer but clearly indicate where it should go.

## Efficient cutting to minimise waste

If a number of components are to be made from the same sheet of metal, thought should be given to how they could be arranged to minimise waste. This is known as **tessellation**.

> **Tessellation**: the arrangement of components to minimise waste.

# 2.7 Specialist techniques, tools, equipment and processes

## 2.7.1 Tools and equipment

REVISED

- **Hand tools**: there are many hand tools that are specific to working metal. Metal tools are quite robust and relatively inexpensive; however, they do require a lot of effort and a great deal of skill to use accurately.
- **Machinery**: powered machines take the physical effort out of working metal; however, they can be expensive to buy and run.
- **Digital design and manufacture**: CAM equipment such as CNC lathes and CNC milling machines use CAD to manufacture metal products quickly, accurately and consistently, with little human input.

## 2.7.2 Shaping

REVISED

### Saws and sawing

- The hacksaw is the most commonly used saw for cutting metal in straight lines.
- A junior hacksaw is a smaller version used to cut smaller sections of metal.

### Shears and shearing

Sheet metal can be sheared using tin snips or a guillotine. Shearing has the advantage of producing a much cleaner cut, but it is limited to thin sections.

### Files and filing

- Once the metal has been roughly sawn it can then be shaped using a file. Several different types of file are used to produce different shapes.
- There are two main filing techniques:
  - **Cross filing** removes the most metal and helps to ensure a flat, level surface.
  - **Draw filing** produces the smoothest surface.
- The quality of the surface finish is determined by the roughness of the file. A 'rough cut' file will remove lots of metal but will leave a rough surface, a 'second cut' file will leave a smoother finish, while a 'smooth cut' file will leave the smoothest finish.

Figure 2.10 **A pair of tin snips shearing through thin sheet metal**

**Figure 2.11** Cross filing

**Figure 2.12** Draw filing

## Drills and drilling

A drill bit will produce a circular hole in a metal component. Before a hole can be drilled, it must be marked out and centre punched to prevent the drill bit from slipping. It is important to secure the metal component before drilling, and you must remember to wear the correct personal protective equipment (PPE).

## Turning

A centre lathe is used to **turn** metal components. The metal is held in a chuck and is rotated while lathe tools perform a number of different operations to shape the component.

## Milling

A **milling** machine is used to cut slots and grooves, and to machine edges and smooth large surface areas of metal.

> **Turning**: a method of producing cylindrical metal components using a lathe.
>
> **Milling**: a method of slotting, grooving and flattening metal using a milling machine.

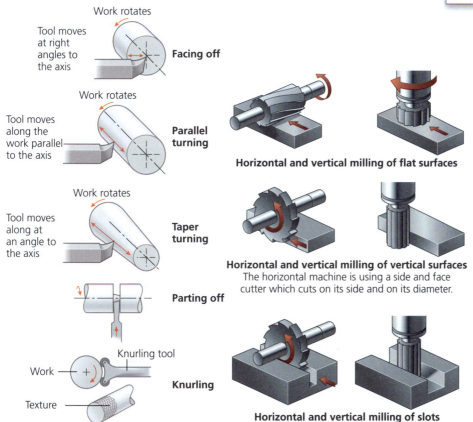

**Figure 2.13** Lathe operations

**Figure 2.14** Milling operations

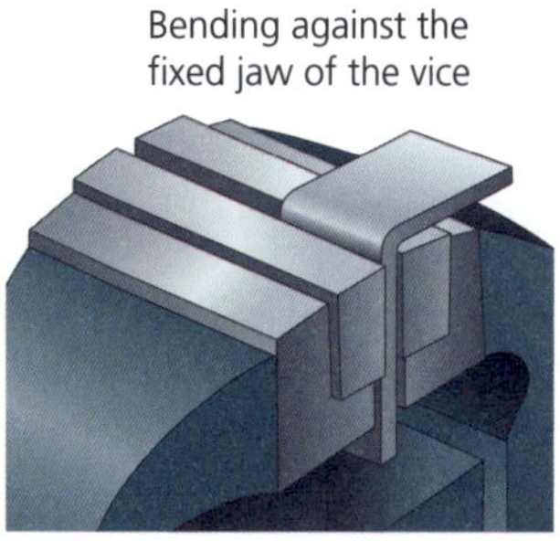

**Figure 2.15** A simple bend in a metalwork vice

## Bending

Metal can be bent by simply holding it in a vice and hitting it with a hammer. For more accurate, longer bends, a metal press or folding bars can be used.

## Grinding

Metal can be ground into shape using an angle grinder. Care must be taken when using hand-held powered machinery as this can be very dangerous. Always ensure that you use the correct PPE.

## Abrading

**Abrading** is the process by which metal can be made smooth using varying grades of abrasive paper.

- Emery cloth has a cloth backing, which makes it durable.
- Wet and dry paper is much smoother and produces a finer finish.

> **Abrading**: smoothing the surface of a material with abrasive papers.

## Casting

See Casting under Section 2.6.1 Processes, on page 60.

## Deforming and reforming

Sheet metal can be deformed and reformed into curved shapes by shaping with a hammer or mallet.

- Hollowing involves hammering sheet metal into a sand-filled leather bag with a bossing mallet.
- To sink a piece of sheet metal it is beaten against a solid edge.
- Sheet metal can be raised to form a cup by beating it against a metal stake.

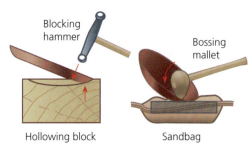

Figure 2.16 **Hollowing a sheet of copper**

# 2.7.3 Fabricating/constructing

REVISED

## Welding

See Welding under Section 2.6.1 Processes, on page 61.

## Brazing

Brazing is a permanent method of joining steel components together. The brazing process is as follows:

1 Firstly, clean the metal surface.
2 Apply a flux (borax) to the joint.
3 Heat the metal until it is cherry red.
4 Apply the brazing rod (brass).
5 Allow the joint to cool.

## Soldering

Soldering is a permanent method of joining softer metals, such as copper, together. It is frequently used by plumbers to join copper piping. The soldering process is as follows:

1 Clean the copper.
2 Apply flux to the joint.
3 Heat the joint until red.
4 Apply the solder (tin).
5 Allow the joint to cool.

## Stamping and punching

See Stamping under Section 2.6.1 Processes, on page 61.

## Riveting

- Riveting is a permanent method of joining metal components together.
- Snap rivets have a domed head and can be made from steel, aluminium or copper.
- Pop riveting is a simpler method of riveting and is useful when you only have access to one side of the joint.

## Sheet metalwork

See Deforming and reforming in Section 2.7.2 Shaping, on the previous page.

## Wasting

Wasting involves cutting and shaping techniques such as sawing, shearing, filing and grinding to produce a metal product.

## Addition

Addition involves joining techniques such as bolting, riveting, welding and soldering to form a metal product.

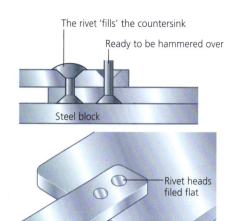

**Figure 2.17 The snap riveting process**

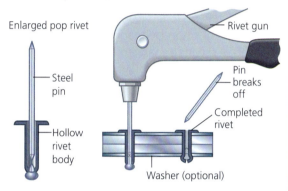

**Figure 2.18 The pop riveting process**

## 2.7.4 Assembling

REVISED

## Tapping

**Tapping** is the process of forming a screwthread inside a hole. A bolt can then be screwed into the hole. The tapping process is as follows:

1 A tapping hole is drilled into the metal.
2 A taper tap is then placed in the hole and turned in a clockwise direction.
3 The taper tap will begin to cut a thread into the hole.
4 After each clockwise rotation the tap should be turned anticlockwise to break off the cuttings (swarf).
5 Applying cutting paste makes the process smoother and produces a higher quality thread.
6 The process is repeated with the plug and bottoming tap to produce a full thread throughout the hole.

> **Tapping**: a method of producing internal screwthreads.

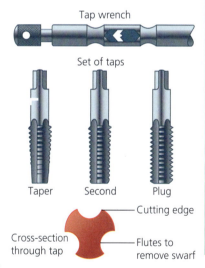

**Figure 2.19 Screwcutting taps and a tap wrench**

## Threading

**Threading** is the process of forming a thread around a metal bar. A nut can then be screwed on to the bar. The process is as follows:

1 The end of a metal bar is chamfered with a file.
2 A screwcutting die is placed into a die holder and rotated clockwise on to the bar.
3 The thread will begin to form on the bar.
4 After each clockwise rotation the die should be turned anticlockwise to break off the cuttings (swarf).
5 Applying cutting paste makes the process smoother and produces a higher quality thread.

**Threading**: a method of producing external screwthreads.

Figure 2.20 **Screwcutting die and die stock**

## Nuts, bolts and washers

● Nuts screw on to a thread bar or a bolt and form one half of a non-permanent joint. They are usually tightened with a spanner.
● Washers protect the surfaces being joined and spread the force of the joint.

## Machine screws

Machine screws screw directly into a pre-tapped hole and hold metal components together.

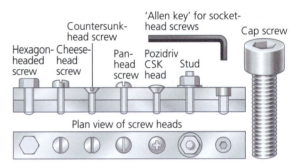

Figure 2.21 **Different types of machine screw**

## Adhesives

Adhesives can be used to glue metal components together.

● **Contact adhesive** is a medium-strength adhesive that glues on contact. To apply contact adhesive:
1 Ensure both surfaces are clean and free from dirt or grease.
2 Apply a thin layer of contact adhesive to both surfaces.
3 Allow to dry.
4 Firmly press both surfaces together.

● **Epoxy resin** is a strong, two-part adhesive. To apply epoxy resin:
1 Ensure both surfaces are clean and free from dirt or grease.
2 Mix the adhesive and the hardener together in equal amounts.
3 Apply to surfaces to be joined.
4 Firmly press and hold both surfaces together until dry.

**Typical mistake**

Adhesives for metals have unique methods of application. Candidates often just deal with applying the glue and lose marks for not addressing the preparation stages.

**Exam tip**

If a question asks for both notes and sketches, be sure to add both notes/labels and diagrams/drawings.

## Now test yourself

TESTED ☐

1 Explain the difference between cross filing and draw filing. (2)
2 What safety precautions would you take when drilling? (3)
3 What name is given to an abrasive paper used for metal? (1)
4 Describe the process of pop riveting. (3)
5 What is the function of a washer when using a nut and bolt? (2)

# 2.8 Surface treatments and finishes

**Table 2.8** Surface finishes and treatments for metals

| Surface treatment | Use | Description | Application | Advantages | Disadvantages |
|---|---|---|---|---|---|
| Paint | Car bodywork, large steel installations such suspension bridges | Paint provides a coloured, protective chemical coating over the metal | Paint can be brushed, rolled or sprayed on to metal | Easy to apply, doesn't require sophisticated equipment, cost effective | Not as durable as some other finishes |
| **Dip coating** | Tool handles | Dip coating gives a polymer barrier over the metal | The metal is cleaned and placed into an oven at 230 °C; then dipped into a fluidising bath containing powdered plastic | Even coating, covers all the surface area, little waste, relatively quick | Specialist equipment is needed |
| **Electroplating** | Jewellery | Electroplating gives a decorative metal coating over a base metal | The metal is chemically cleaned then suspended in an electrolytic solution; a weak electrical charge flows between the base metal and the coating metal | Even coating, provides a decorative finish | Specialist equipment is required, relatively expensive process |
| **Anodising** | Specialised aluminium components, such as parts for a camera, mountain bike pedals and levers | A coloured oxide coating for use on aluminium | The metal is chemically cleaned then suspended in an electrolytic solution; a weak electrical charge flows between the base metal and an anodic dye | Very durable, available in exotic metallic colours, very corrosion resistant, hardens the surface of the aluminium | Specialist equipment is required, relatively expensive process |
| **Galvanising** | Barriers on a motorway, car body panels | A coating of zinc is applied over steel | Steel is dipped into molten zinc | Very corrosion resistant, very durable | Gives a dull grey finish, can only be used on steel |
| **Powder coating** | White goods such as fridges, freezers, cookers, washing machines | A thin polymer coating is applied over a base metal | The metal is cleaned then coated with a powdered polymer via an electrostatic gun; then cured in an oven | Provides a thin protective coating, very durable, very corrosion resistant, very little waste | Specialist equipment is required |

➡️

**Table 2.8** Surface finishes and treatments for metals (continued)

| Surface treatment | Use | Description | Application | Advantages | Disadvantages |
|---|---|---|---|---|---|
| Lacquering | Jewellery | A thin layer of clear cellulose lacquer covers surface of the metal | The metal is cleaned and then sprayed, brushed or rolled with lacquer | Gives an even, thin, transparent coating, protects the metal from tarnishing | Not very durable |
| Polishing | Jewellery | The surface of the metal is polished to produce a shiny finish | A polish is applied to the metal then rubbed or buffed | Gives a high-quality shiny appearance | Not very durable, needs to be repeated regularly |

**Dip coating**: coating metal with a polymer using heat.

**Electroplating**: coating one metal with another metal using an electrolytic process.

**Anodising**: coating aluminium with an anodic dye using an electrolytic process.

**Galvanising**: coating steel with zinc using a hot dip method.

**Powder coating**: coating metal with a polymer by electrostatic spraying then heating.

**Exam tip**

Make sure that you can match a metal finishing process with the correct metal and provide examples of where it could be used.

## Now test yourself

TESTED ☐

1 Why must a steel product be given a finish? (1)
2 Name the metal that you would anodise. (1)
3 What are the advantages of galvanising car bodies? (2)
4 Describe the process of powder coating. (3)
5 What are the disadvantages of painting with a brush? (3)

**Typical mistake**

When describing a metal finishing process, many candidates lose marks by not addressing surface preparation.

## Exam practice

1 Give a property of aluminium that makes it suitable for use in an aircraft. (1)
2 Give a property of steel that makes it suitable for use in a car body. (1)
3 Give **two** reasons why metal can be considered a sustainable material. (2)
4 Explain why nuts and bolts are bought as standard components. (3)
5 What is meant by the term 'stock form'? (2)
6 Use notes and sketches to describe the process of tempering. (3)
7 Describe the process of tapping a hole. (6)
8 Describe the process of sintering (powdered metallurgy). (4)
9 Explain the advantages and disadvantages of continuous production. (4)
10 Describe the process of anodising. (4)

# 3 Papers and boards

## 3.1 Design contexts

- The **context** of a design incorporates such things as:
  - the surroundings or environment where it will be used
  - the different **users** and **stakeholders**
  - the purpose of the end product
  - social, cultural, moral and environmental considerations.
- A product designed within context will fulfil its purpose exactly, giving the users and stakeholders what they require.
- If design takes place without considering the context, the final product will not fully meet the needs of the user or stakeholders.

> **Typical mistake**
>
> Although the user may often appear to be the only stakeholder, this is rarely the case. Anyone who may come into contact with a finished product can be considered a stakeholder.

> **Design context**: the settings or surroundings in which the final product will be used.
>
> **User**: the person or group of people a product is designed for.
>
> **Stakeholder**: a person other than the main user who comes into contact with or has an interest in the product.

> **Exam tip**
>
> Toys are products designed for children, but parents purchase the toy and have to store, carry, clean and maintain it. Although the child is the main user, the parents are a major stakeholder and their needs must also be considered by the designer.

## 3.2 Sources, origins, physical and working properties, and social and ecological footprint

### 3.2.1 Paper

REVISED ☐

Copier paper, cartridge paper and tracing paper are the most common types of paper. These are discussed in Section 1.9.1 Paper on page 28.

### Bond paper

- Bond paper is a durable, high-quality paper that was originally used for important documents such as government 'bonds'. It is still used for many legal documents.
- Bond paper has a weight of between 50 gsm and 100 gsm.
- Cotton rag fibres are often incorporated, which makes the paper much stronger but rougher.
- Bond paper is more expensive than ordinary paper.

### Heat transfer paper

- Heat transfer paper is used to transfer printed designs on to T-shirts, mugs and other products.
- A design is created on a computer and printed on to the paper using an inkjet or laser printer.
- The printed design is placed face down on the material and pressed with an iron or heat press.

- This releases the ink into the material, where it permeates the fibres then solidifies.
- Heat transfer paper is quite expensive and only suitable for small printing runs or one-offs.

## 3.2.2 Board

Folding boxboard, corrugated cardboard and solid white board are the most common types of board. These were covered in Section 1.9.2 Board on page 29.

### Foil-lined board

- Foil-lined board is thin card with a laminated foil surface.
- The foil layer makes the surface waterproof and helps to retain heat.
- Foil-lined board is used for drink cartons and food packaging.

## 3.2.3 Packaging laminates

- Packaging laminates are made up of multiple layers of paperboard, polyethylene and aluminium foil.
- The layers used depend on the use and properties required.

### Paperboard

- Paperboard is stronger and thicker than normal paper; it has a smooth surface that is good for printing.
- It is used for paperback book covers, postcards and other packaging.
- Unlike other packaging laminates, paperboard is not waterproof.

### Polyethylene

- Polyethylene is a flexible polymer layered with foil to make it waterproof and retain heat.
- It is often used in sealed food packages to retain preservative gas inside.
- It takes a long time to biodegrade and the multiple layers are difficult to separate for recycling.

### Aluminium foil

- Aluminium foil is a flexible and easily shaped laminate that is waterproof and totally light resistant.
- It is used in food packaging and other cooking and kitchen applications.
- It is quite fragile and easily torn or punctured compared to other laminates.

### Tetra Pak

- **Tetra Pak** is made up of six layers – four layers of polythene with a layer of paperboard and foil.
- Tetra Pak can keep food well preserved for long periods of time without refrigeration and is used for long-life food and drink packaging.
- It is expensive to manufacture and difficult to recycle due to its multiple layers.

> **Tetra Pak**: laminate made up of layers of polythene, paperboard and foil; it is used for long-life food and drink packaging.

## 3.2.4 Sources and origins

- 35 per cent of the total trees felled around the world are used in paper making.
- The majority of trees used are from the USA, Japan, China, Canada, Germany and Finland.
- Rice paper is made in East Asia from rice straw and other plants, such as hemp and bamboo.

## 3.2.5 Physical properties

The main physical characteristics of paper and board are:

- density – calculated by dividing its mass by its volume
- **transparency** – the amount of light transmitted through the paper
- **texture** – the feel and finish of the paper (smooth or rough).

**Transparency**: the amount of light transmitted through a material.

**Texture**: the feel and finish of a material.

## 3.2.6 Working properties

The flexibility, printability and biodegradability of paper and boards were covered in Section 1.9.3 Properties on page 30. The other main working properties are:

- weight – measured in **grams per square metre (gsm)**
- surface finish – the look and feel of the surface (for example, shine, colour, smoothness)
- absorbency – the amount of moisture or water that can be absorbed.

**Grams per square metre**: the weight in grams of one square metre of paper.

## 3.2.7 Social footprint

### Trend forecasting

- **Trend forecasting** means predicting the upcoming colours, textures, materials and graphics that will be in demand in the near future.
- Trend forecasting is used to help attract and retain customers.

**Trend forecasting**: predicting the upcoming colours, textures, materials and graphics that will be in demand in the near future.

### Impact of material production and logging

- Harvesting trees to make paper and board can cause soil erosion, so crops cannot grow, and creates water pollution.
- Vehicles transport timber to paper processing plants – this releases carbon dioxide and other emissions into the air.
- Processing the timber into paper uses lots of energy and water, which can affect the water supply for local communities and wildlife. Processing plants also use heavy machinery, which can create noise and air pollution, and dust, which can cause respiratory problems.
- Polluted water from the paper making process can affect rivers, while waste pulp fibre is either burned or goes into landfill.

### Ethical responsibility

- Manufacturers of paper have an ethical responsibility to be as environmentally friendly as possible.
- Many consumers refuse to buy products from companies that are not environmentally friendly.

- Symbols on products, such as the EU Ecolabel, show consumers that the manufacturer uses eco-friendly materials and processes.
- It is important for products to be recycled or disposed of correctly – see Section 1.14.4 Recycling and reusing materials or products on page 40.

## Reduction of packaging materials

- Many products have excess packaging, which takes more energy to produce, causes litter and is often sent to landfill.
- Using less packaging, or smarter packaging such as recyclable materials, can give the same amount of protection but cut down on energy use, waste and litter.

## Brand identity

- **Brand identity** is when manufacturers use specific colours, fonts, shapes, designs and logos to distinguish themselves from other brands. This helps consumers identify the brand quickly and easily when purchasing products.
- **Consumerism** is when a business responds to the needs and requests of its customers in a certain way. This influences the way that products are manufactured, transported or packaged.
- Many businesses have changed their packaging in response to requests from customers to be more environmentally friendly. This makes the customer more likely to remain loyal to the brand.

> **Brand identity**: the use of specific colours, fonts, shapes, designs and logos by manufacturers to distinguish themselves from other brands and portray the right image to consumers.
>
> **Consumerism**: a business responding to the needs and requests of its customers in a certain way.

## 3.2.8 Ecological footprint

REVISED

- Although paper and boards are considered to be sustainable materials, the level of consumption has meant that trees are being cut down and used faster than they can be regrown. If this trend continues, the number of trees will decline, resulting in widespread deforestation.
- Managed forests plant new trees for each one that is felled and use harvesting methods that 'thin' the forest by felling small trees every few years.
- Managed forests help to preserve the habitats of wildlife and prevent erosion of the natural landscape.
- Wood processing plants create dust, water pollution and noise. However, they can also provide employment, improving the local economy and benefiting the local community.
- Transportation of trees from forests to processing plants uses fossil fuels, which produces carbon dioxide and contributes to global warming.
- Waste paper fibres that cannot be recycled often end up in landfill, but they can be dried out and burnt as a fuel.
- The paper-making process uses solvents, bleaches and other toxic chemicals that can pollute the air and water.

> **Exam tip**
>
> The Forest Stewardship Council® (FSC®) manages forests where new trees are planted for every tree felled.

> **Typical mistake**
>
> Paper and boards are considered to be sustainable materials as they are made from trees which can be regrown.
>
> However, hardwoods (which take a long time to grow) are in decline as they are being cut down and used faster than they can be regrown. These are much less sustainable than softwoods.

## Now test yourself

TESTED

1 What type of paper would you use to print a design onto a T-shirt? (1)
2 What is bond paper? (2)
3 Why can paper only be recycled five or six times? (3)

# 3.3 Selection of papers and boards

## 3.3.1 Aesthetic factors

REVISED

Aesthetic factors relate to how the product looks and feels:

- form – the size, thickness and physical properties of the paper or board
- colour – the colour, shade and finish of the paper or board
- texture – the surface feel of the paper or board
- surface graphics – the ability of the paper or board to be drawn, painted, written and printed on using different media.

## 3.3.2 Environmental factors

REVISED

- Sustainability – the ability to regrow the materials needed to make paper and board.
- Pollution – the amount of damage done to the environment during the paper-making process, for example by the chemicals used and carbon dioxide emissions.
- **Genetic engineering** – manipulation of the DNA of plants used in the making of paper and board to improve their physical properties.

> **Genetic engineering**: manipulation of the DNA of plants used in the making of paper and board to improve their physical properties.

## 3.3.3 Availability factors

REVISED

- Stock materials – widely available 'off-the-shelf' paper and board materials with specific and consistent properties that are cost efficient.
- Specialist materials – more expensive and difficult to source papers and boards that may need special handling, storage and manufacturing processes but result in a higher quality product.

## 3.3.4 Cost factors

REVISED

Designers should be able to calculate costs and work within the cost constraints of the client's brief, considering:

- quality – the cheapest products are usually of a lower quality
- decorative techniques – embossing, varnishing and hot foil blocking will increase costs
- manufacturing processes necessary – high-quality finishes such as super calendaring and full-colour printing will increase the costs
- **commodity price** – the cost of raw materials needed to make the paper or board, which can go up or down in price and affect the unit cost
- cost of recycling – it is often cheaper to use recycled materials than new raw materials; certain products are difficult and expensive to recycle, however, which makes it cheaper to use 'new' raw materials.

> **Commodity price**: the cost of raw materials.

> **Exam tip**
>
> It is often cheaper to use recycled materials instead of virgin raw materials to make new products, as less new raw materials need to be purchased. Certain products are difficult and expensive to recycle, however, which makes it cheaper to use 'new' raw materials.

## 3.3.5 Social factors

REVISED

- Social groups – groups of people may have different feelings about the use of certain types of paper and board. Some groups object strongly to the amount of unnecessary packaging used by some supermarkets.

- Trends and fashion can influence the materials and finishes used so that the product will appeal to the current market.
- The popularity of other products can influence designers – they may try to emulate or produce products with similar attributes and properties using similar materials.

## 3.3.6 Cultural and ethical factors

REVISED

- Avoiding offence – offence can be caused by using the wrong wording, images or colours on products.
- Suitability for intended market – matching the product to the intended market will ensure the product is successful.
- Use of colour and language – colours and words can have different meanings and connotations in different parts of the world.
- **Consumer society** – people feel a need to have the most up-to-date products; they are used as a display of a person's status.
- Effects of mass production – high-volume production can mean that the supply of some products is much greater than the demand for them. Manufacturers are then left with unwanted stock.
- **Planned/built-in product obsolescence** – when products are deliberately made to only last a short period of time, so that consumers need to buy new products from the manufacturer more frequently.

> **Consumer society**: the need for people to have the most up-to-date products.
>
> **Planned/built-in obsolescence**: when a product is designed to no longer function or be less fashionable after a certain period of time.

> **Typical mistake**
>
> Obsolescence is not always planned by the manufacturer or designer. Products can become obsolete through the introduction of new technological advances and discoveries.

> **Now test yourself**
> TESTED
>
> 1  What **three** social factors can influence a designer's selection of paper and boards? (3)
> 2  What is meant by the term 'aesthetics'? (2)
> 3  What is the term given to the desire people feel to have the most up-to-date products, even if they do not actually need them? (1)

# 3.4 Forces and stresses

## 3.4.1 Forces and stresses

REVISED

Forces are applied to papers and boards in different ways, as shown in Table 3.1.

**Table 3.1** Forces and stresses acting on papers and boards

| Type of force | Description |
| --- | --- |
| Bending | Forces acting at an angle to a material |
| Torsion | Twisting forces on a material |
| Shear | Forces acting across a material |
| Compression | Pushing forces pressing on to a material |

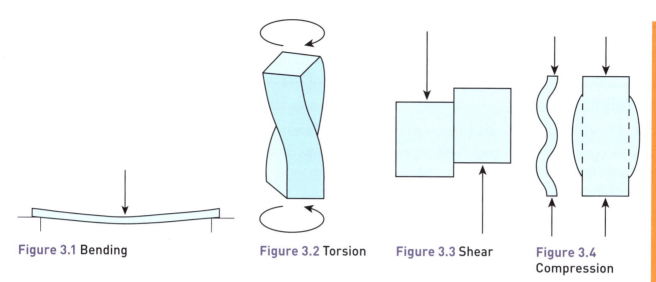

**Figure 3.1** Bending      **Figure 3.2** Torsion      **Figure 3.3** Shear      **Figure 3.4** Compression

**Bending force**: a force acting at an angle to a material.

**Torsion force**: a twisting force on a material.

**Shear force**: a force acting in opposite directions across a material.

**Compression force**: a pushing force pressing on to a material.

> **Exam tip**
>
> Paper is far stronger in tension than in compression.

## 3.4.2 Reinforcement/stiffening techniques

REVISED

- **Laminating** – covering the paper or card with a layer of thin plastic to make it stronger, less likely to tear and more resistant to moisture.
- **Encapsulation** – the paper or board is fully enclosed within a heat-sealed plastic pouch. This makes it completely waterproof.
- **Corrugation** – this improves the **compressive strength** of card. A wavy central layer creates a triangulated structure that spreads any force across the waves of triangles. Commonly used for packaging products.
- **Additions of layers** – adding extra layers increases the thickness and rigidity of papers and boards.
- **Ribs** – thin lengths of card or other materials that are added to paper or card structures in specific places to add strength and rigidity without adding unwanted weight.
- **Sandwich construction** – sandwiching other lightweight materials between layers of paper and board can significantly increase strength and rigidity without adding weight, e.g. foam board, where a layer of thin foam is sandwiched between two layers of paper or thin card.
- **Packaging laminates** – other materials, such as plastic and foil, are layered with paper in different combinations to create the properties required. Packaging laminates have high strength but remain flexible and are easy to print on to in high definition. The **lamination** makes them resistant to moisture and extreme temperatures.

> **Compressive strength**: the ability to withstand a pushing (squashing) force.
>
> **Lamination**: covering with a layer of thin plastic to make the material stronger and more resistant to moisture.

> **Typical mistake**
>
> Paper is not considered to be a 'strong' material, but when used in certain ways it can be extremely strong.

### Now test yourself

TESTED

1  What is 'sandwich' construction?      (2)
2  What is encapsulation?      (2)

# 3.5 Stock forms, types and sizes

## 3.5.1 Stock forms/types

### Weight

- Paper is classified by weight, measured in grams per square metre (gsm).
- By buying paper in standard weights, the designer can accurately calculate how much is required.
- Paper is sold in **reams**. A ream is 500 sheets.

> **Ream**: a pack of 500 sheets of paper.

### Bond

- During the paper-making process, layers of tiny wood fibres are placed on top of each other and 'bond' by sticking together.
- The internal bond strength (IBS) of papers and boards refers to the strength of the bonding of these fibre layers and their resistance to ripping or tearing.
- The **bond strength** is the force required to separate or rip the paper and can be expressed in Newtons per metre (N/m) or energy per unit surface ($J/m^2$).

> **Bond strength**: the force required to separate or rip paper expressed in Newtons per metre (N/m) or energy per unit surface ($J/m^2$).

### Laminates

- Laminates are special types of coated papers for specific purposes such as the packaging of food and drinks.
- Laminates are supplied in large rolls, up to two or three metres wide, that weigh over a tonne.

## 3.5.2 Sizes

### Common A sizes

- Sizes range from A0 (the largest) down to A10, with the most common size being A4.
- Each paper size is half the area of the one before, for example A4 paper (297 mm × 210 mm) is half the size of A3 paper (297 mm × 420 mm).

> **Exam tip**
>
> Learn the size of A4 paper (297 mm × 210 mm) and use this to calculate the size of other A sizes.

### Foolscap

- Foolscap paper measures 8.5 × 13.5 inches (216 mm × 343 mm), or 13 × 8 inches (330 mm × 200 mm).
- Foolscap was the most commonly used paper size in Europe and the Commonwealth before the international standard A4 paper size was adopted.

> **Typical mistake**
>
> A3 paper is not smaller than A4. The smaller the number, the larger the size.

### B series

- The B series of paper sizes is not as common as the A series. It is used to describe both paper and printing press sizes.
- B5 is a common size used for books, and many posters use B series paper sizes or very close to them, such as 50 cm × 70 cm. The B series is also used for envelopes and passports.

## Letter

Letter size paper is commonly used in the USA and Canada; it is 8.5 inches × 11 inches (216 mm × 279 mm).

## Envelope

- Envelope sizes are designed to fit A series paper sizes.
- C4 envelopes are 229 mm × 324 mm (approximately 20 mm longer and wider than A4) and are designed to fit A4-sized documents or A3 documents folded in half.
- C5 envelopes are 162 mm × 229 mm (approximately 20 mm longer and wider than A5) and are designed to fit A5-sized documents or A4 documents folded in half.
- DL envelopes are 110 mm × 220 mm and are designed to fit A4-sized documents folded twice.

## Area

The area of standard paper sizes can be calculated easily by multiplying the length by the width.

> ### Now test yourself                                    TESTED ☐
>
> 1 Why is bond paper used for legal documents?            (1)
> 2 Explain the relationship between A4 and A3 size paper.  (2)
> 3 What is a 'ream' of paper?                              (1)

# 3.6 Manufacturing to different scales of production

## 3.6.1 Processes                                    REVISED ☐

### Digital printing

- The most cost-effective production method for small print runs.
- It allows variations such as enlargement and reduction, cropping, rotating and personalisation.
- Digital printers are inexpensive to buy.
- The **cost per sheet** is high compared to other types of commercial printing.

### Screen printing

**Screen printing** is used for creating repeating patterns or designs such as wallpaper or fabrics. The process is as follows:

1 A porous fabric mesh screen is stretched over a wooden frame called the screen.
2 The mesh screen is masked up for the first (lightest) colour.
3 The frame is laid on to the paper or fabric, and ink is pushed through the mesh with a squeegee. The screen is then moved to the next position.

> **Screen printing:** the process of creating a design by pressing ink through a stencilled mesh.

4 Once the lightest colour has been printed, the screen is washed and masked up for the next colour. The process is repeated until all the colours have been printed.

## Offset lithography

- Offset lithography is used for large print runs.
- The process uses four ink colours: cyan, magenta, yellow and black (called the 'key'). This is often shortened to CMYK.
- The four colours are overlaid to create others; for example, printing cyan on top of yellow creates green.

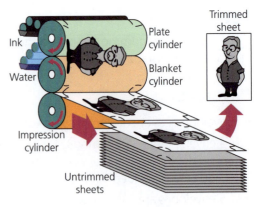

**Figure 3.5 Offset lithography**

## Flexography

- Flexography is a mass production printing process. It uses water-based inks that dry quickly but the print quality is not as good.
- It is largely used for printing on packaging, such as corrugated cardboard boxes and plastic carrier bags, where the quality of print is not as important.

## Cutting

Scissors and craft knives are the most obvious choice for cutting paper and card. Other types of cutting equipment include:

- guillotine (paper trimmer) – used to cut a large number of paper sheets at once with a straight edge
- compass cutter – used to cut a circle or an arc from paper or thin card
- rotary cutter – used to cut a circle or an arc from thicker cardboard
- die cutter – used to cut, crease and perforate paper and card in large numbers in commercial use
- laser cutter – used to cut any 2D shape out of card, PVC foam, foam board or Corriflute.

For more on hand tools used for cutting, see Section 3.7.1 Tools and equipment on page 83.

## Intermediate modelling of paper and card prototypes

- Paper and board are extremely versatile materials for model making as they can be folded, cut, bent, printed on and stuck together easily.
- Printed parts or patterns can be cut out and assembled by slotting or gluing.
- Nets can be drawn or printed, then scored, folded, cut out and joined to form the shape required.
- Foam board is often used to make architectural models as the thickness of the material means it can be stood up and joined to form walls, floors and roofs.

## Frame modelling

- Frame models are often used in construction projects for buildings or bridges.
- Frame layouts are drawn out on paper and the frames are put together over the drawing.

> **Frame modelling:** joining strips of thin softwood or balsa by gluing triangular-shaped card pieces called gussets over each joint to create a rigid frame.

## Test modelling

**Test modelling** is modelling a part of a design to see if it will work as intended. Models made from card are often used to test mechanisms, linkages and other moving parts.

> **Test modelling:** modelling part of a design to see if it will work as intended.

## 3.6.2 Scales of production

REVISED

- **One-off production** – paper and boards are ideal for the manufacture of one-off prototypes due to the low cost and ease of manipulation of the material.
- **Batch production** – promotional books and magazines, such as concert or football match programmes, are printed in batches as they are needed in a reasonably large quantity but only printed once.
- **Mass production** – used for very large quantities, such as printing popular magazines and newspapers.
- **Continuous production** – when the product is in constant production (24 hours per day, seven days per week) as there is a continual demand for it. An example is newsprint, the paper used for printing newspapers onto. The process is fully automated and uses highly specialised equipment, but unit costs are low.

## 3.6.3 Techniques for quantity production

REVISED

### Marking out methods

- Quality-control procedures are used when printing products to ensure a high-quality print.
- Registration marks ensure the printing lines up correctly on the pages.
- Crop marks show where to cut the pages.
- Colour bars are printed on the edges of the page to ensure colours are being printed to the correct consistency and intensity.

### Fixtures and folding jigs

- Fixtures allow card and board products to be assembled accurately.
- Folding jigs are fixtures that allow cardboard box nets to be assembled quickly and accurately by hand. The pre-cut card nets are laid on to the jig and pushed down into the frame that forms the box shape.

### Templates, stencils and patterns

- Stencils, patterns and templates can be used for batch production. Where multiple numbers of the same part are required, it can be drawn out once on card and cut out to create a template.
- The template or pattern can then be laid on to the material required and drawn around, then moved and drawn around again until the required number is reached.
- Stencils have the shape or design cut out from a piece of card, which is then placed on to the material required. The inside edge of the shape(s) is then sprayed or drawn around.
- Using a template, pattern or stencil ensures that every piece is exactly the same and saves the time of drawing each part out individually.

> **Typical mistake**
>
> Stencils are not just used for lettering. They can be used for repeating shapes and designs on to other materials such as textiles, timber and metal.

## Photocopying

- Photocopying is a quick and cheap way to make copies of documents or images on paper or plastic film.
- Higher quality photocopiers that can print and copy in high volumes are used by print and design companies.

## Computer-aided manufacture

- Printers are the simplest and most common types of computer-aided manufacture (CAM) machines used for transferring on-screen text and images to paper or board.
- Vinyl cutters allow designs to be cut out in self-adhesive vinyl, which can be stuck on to other media such as vehicles and signs. Vinyl cutters don't require specialised or expensive software to operate.
- Laser cutters can cut 2D shapes by burning through the material; this can leave scorch marks on the edge of the material, however.

> **Exam tip**
>
> Printers and plotters are types of CAM machinery.

## Quality control

- Quality control ensures that products being manufactured meet various standards and are free from defects.
- The product is checked at various stages in the manufacturing process to ensure it is within pre-set limits of size, quality, finish, and so on.
- Items that do not meet quality-control standards are rejected.

## Tolerance

- **Tolerance** is the allowable amount of error or variation when manufacturing a product.
- The tolerance can relate to things such as dimensions, measured values or physical properties of the material.
- A variation outside the tolerance, such as a measured length that is too big, is 'out of tolerance' and unacceptable.
- For example, if a sheet of card needs to be 1000 mm long with a tolerance of ± 1 mm, the card must be between 999 mm and 1001 mm long to be 'in tolerance'.

> **Tolerance**: the allowable amount of variation of a stated measurement.

## Efficient cutting to minimise waste

**Tessellation** should be used to reduce waste when cutting out. Tessellation is when objects are arranged in such a way that the material being cut out is used to its maximum capacity and there is only the minimum amount of waste.

> **Tessellation**: the arrangement of components to minimise waste.

### Now test yourself

TESTED

1 What will be the longest and shortest acceptable lengths for a strip of card that is 450 mm long with a tolerance of ± 2 mm? (2)
2 What is tessellation? (2)
3 Describe **one** advantage and **one** disadvantage of continuous production. (4)

# 3.7 Specialist techniques, tools, equipment and processes

## 3.7.1 Tools and equipment

### Hand tools

- Scissors and craft knives can be used for cutting paper and card.
- Craft knives can be used to cut foam board, PVC foam, Corriflute and Styrofoam™ up to a thickness of around 10 mm.
- A serrated knife (such as a bread knife), bandsaw, hacksaw blade or hot wire cutters can be used to cut Styrofoam™ sheet.

### Machinery

- Paper cutters (also called paper trimmers or guillotines) can be used to cut large sheets of paper and cardboard accurately to size.
- Paper cutters cut along the entire edge of the sheet in a straight line.

### Digital design and manufacture

- Digital design tools are computer programs that allow designers to create 2D drawings and 3D models of designs.
- Computer-aided design (CAD) software can be used to draw up complete designs or parts of designs that can be sent to CAM machines such as vinyl cutters and laser cutters. This allows models and mock-ups of designs to be modelled quickly and accurately.

## 3.7.2 Shaping

### Cutting

- For intricate shapes in paper and card, scissors and craft knives are the best choice – see Section 3.7.1 Tools and equipment above.
- A safety rule allows the safe cutting of straight lines.
- Compass cutters are like a compass but have a blade instead of a pencil and are used to cut a circle or an arc from paper or thin card.

### Folding

- Paper and thin card can be folded easily by hand.
- Scoring the material first using a blunt knife blade or other dull, pointed object will help fold thicker card and ensure a clean, sharp crease.
- Foam board can be folded by cutting through the foam using **hinge cutting** or **vee cutting**.
- PVC foam cannot be folded unless it is cut partway through.
- Corriflute is not easily folded, but it can be done by cutting a section of material away from the top layer between the flutes.

### Notching

- Notches are shapes made in paper or board that make the material easier to fold, bend or slot together.

> **Hinge cutting**: the foam board is cut part-way through so that the bottom layer of card acts as a hinge and the card can be folded backwards.
>
> **Vee cutting**: a V-shaped cut made in foam board that allows the board to be folded inwards; it gives a clean, tidy fold.

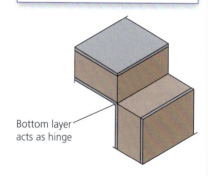

Bottom layer acts as hinge

**Figure 3.6** Hinge cutting

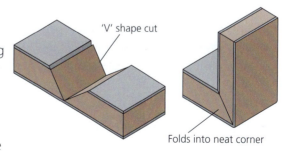

'V' shape cut

Folds into neat corner

**Figure 3.7** Vee cutting

- A shaped punch presses down and cuts the required shape into the board.
- Notching machines are used in mass-produced products.
- A craft knife can be used to create notches for one-off products.

## 3.7.3 Fabricating/assembling/constructing

### Strengthening

- Adding extra layers increases thickness and rigidity.
- Thin lengths of card or other materials such as balsa wood are stuck on to sections of paper and card structures to add strength and rigidity without adding unwanted weight.
- Sandwiching other lightweight materials between layers of paper and board can significantly increase the strength and rigidity – see Section 3.4.2 Reinforcement/stiffening techniques on page 77.

### Addition of dissimilar materials

- Clear plastic windows are used in card and board products, such as packaging, to allow the product inside to be seen by the consumer.
- Thin, clear plastic sheet called acetate is cut slightly larger than the window and glued around the edge.
- Stickers are a quick way of applying information, colour or a design.
- Inserts are extra layers of paper or board added inside a product, such as a greetings card, to add extra detail.

### Lamination

See Section 3.8 Surface treatments and finishes on page 86.

### Split pins

- Split pins are used for joining multiple sheets of paper or board together. The pins are inserted into a hole in the material, then the legs are separated and bent over to secure the paper and board.
- Split pins are often used in modelling for mechanisms as they hold the pieces together but still allow them to rotate.

### Mapping pins

- Map pins are a type of push pin used to mark places on maps.
- They are also used to fasten items to a wall or board for display. They are easier to remove than drawing pins due to the large plastic head.

### Stapling

- Staplers are used to attach sheets of paper and board by forcing a thin metal staple through them and folding the ends to hold them in position.
- Staple guns are used to fasten paper and board to a solid backing.

### Taping

- Adhesive tapes can be used in place of mechanical fasteners, simplifying the process of prototyping.
- Different types of tapes include masking tape, sticky tape, parcel tape and double-sided tape.

# Paper engineering

- **Paper engineering** is the technique of cutting and folding paper into complex 3D shapes, structures and mechanisms.
- Sheet paper and board models can then be reproduced in other sheet materials to create a working prototype.
- Products such as car airbags and solar panels are modelled using paper engineering techniques.

> **Paper engineering**: the technique of cutting and folding paper into complex 3D shapes, structures and mechanisms.

# Use of adhesives

- **Paper and thin card** – glue sticks are an easy method of gluing paper together. They have a bond that is only strong enough for paper and very thin card. Spray mount (spray adhesive) comes in an aerosol can and is a quick way of mounting drawings on to backing paper. It gives a thin, even coat that dries quickly.
- **Boards, corrugated cardboard** – PVA glue is ideal for joining thicker card and corrugated card. Superglue and hot glue guns can also be used when a fast-setting bond is required.
- **Foam board and Styrofoam™** – PVA glue is also ideal for these materials. Masking tape or dressmaker's pins can hold the materials together while the glue dries. Hot glue guns must not be used as the solvent will melt the foam.

> **Exam tip**
>
> PVA glue is commonly used for joining timber and can also be used to join all papers and boards.

> **Typical mistake**
>
> Solvent-based glues, such as superglue, cannot be used on foam board or Styrofoam™ as they melt the material.

# Lettering

- Stencils are an easy way to apply lettering using pens and pencils.
- Lettering can be cut from self-adhesive vinyl using a vinyl cutter.
- Dry transfers are placed over the material then rubbed. They can be used to put high-quality lettering on to card and paper prototypes.

# Binding

- **Ring or comb binding** – holes are punched along the edges of the pages and a spiral ring or plastic comb binder is inserted. Pages can be folded all the way back without damage to the spine.
- **Plastic spines** are a U-shaped length of plastic that can be forced apart so that they grip the pages placed between. They are easy to fit but can become loose over time.
- **Stud binding** (also known as screw or post binding) is a very secure method of joining pages. Holes are drilled through all the pages then a stud is pushed through and an end cap fitted. This requires a wide enough margin on the left-hand edge of the page and means this area of the page cannot be seen after binding.

# Marking out tools

- Pencil or pens can be used to mark out on paper, card and foam board.
- Styrofoam™ can be marked with a thin permanent marker but this can leave a mark on the material.
- A Chinagraph pencil or non-permanent marker can also be used but the lines can be smudged easily.

## Now test yourself

TESTED

1  State **one** advantage of ring binding.                        (2)
2  Name **two** ways of folding foam board.                       (2)
3  Explain why solvent-based adhesives are not suitable for
   Styrofoam™.                                                     (2)

# 3.8 Surface treatments and finishes

## 3.8.1 Surface finishes and treatments

REVISED

### Varnishing

- Varnish can be applied to paper and card products to enhance their look and feel. The coating adds additional protection.
- A varnished card has a high shine finish and feels like plastic.

### Hot foil blocking

- **Hot foil blocking** is used to produce metallic finishes.
- Multi-coloured and holographic foils are also available.
- It is often used for lettering on invitations and business cards.

> **Hot foil blocking**: used to produce metallic finishes on card and paper products.

### Edge staining

- Edge staining is the application of colour to the edges of paper or boards.
- It is done after all text and graphics have been printed and the paper or card material has been cut to its final size.
- It is commonly used on products such as children's books and cards.

### Embossing and debossing

- Embossing and debossing give paper and card a 3D image.
- Embossing creates a raised area on the paper or card that stands out.
- Debossing has the opposite effect and creates a sunken or lowered area.

### UV varnishes

- UV varnishes are applied over ink printed on paper or boards; they dry almost instantly when exposed to UV light.
- UV coatings give the paper or board surface a highly reflective and glossy finish, and deepen colours printed on the page.
- **Spot UV varnish** is only applied on certain areas to make those areas shinier and clearer.

> **Spot UV varnish**: a special varnish dried and hardened by UV light; it is only applied on certain areas to make those areas shinier and clearer.

## Packaging laminates and films

- Laminating involves applying a film of clear plastic between 1.2 mm and 1.8 mm thick to either one or both sides of paper or thin card.

- **Pouch lamination** uses thin clear plastic pouches that are available to fit standard paper sizes. The pouches are coated inside with a thin layer of heat-activated glue. The document is placed inside the pouch and fed through a laminating machine, where it is heated and pressed between rollers, activating the glue, which seals the pouch together.

- **Thermal lamination** uses rolls of thin, heat-sensitive plastic film. Different types of film are available with different properties:

  - BOPP/OPP (polypropylene) film has good resistance to water, chemicals and cracking, excellent transparency, and can be recycled. It is commonly used for write on/wipe off products.

  - PET (polyester) film has all the benefits of BOPP/OPP with improved gloss, scratch resistance and durability. It is commonly used for book covers and documents that have undergone foil application.

  - Nylon film is more expensive due to the high temperatures needed for it to laminate, but it has high chemical, mechanical and abrasion resistance. It absorbs moisture and is commonly used for soft covers such as paperback books.

- **Cold laminating** is done when only one side of the paper or card is to be coated. It is ideal for documents that can be damaged by heat, such as photographs. Cold lamination is mainly used in the sign-making industry and on other graphic products such as foam board.

> **Cold laminating**: coating one side of paper or card without the use of heat.

### Now test yourself

TESTED ☐

1 What is edge staining? (2)
2 Explain the difference between 'normal' varnish and UV varnish. (2)
3 Describe the cold lamination process. (3)

### Exam practice

1 Explain why Tetra Pak is a difficult material to recycle. (5)
2 Explain **two** reasons why corrugated card is often used to make boxes for takeaway pizzas. (4)
3 An A2-sized poster is to be included as a free gift inside a magazine. The magazine is A4 size. How many times will the poster need to be folded in half to fit inside the magazine? (1)
4 Describe the six stages in the lifecycle of a paper coffee cup. (6)
5 Explain why many products are produced with 'planned obsolescence'. (3)
6 Describe **two** surface treatments that can improve the appearance of paper or card products. (2)
7 Using an example, describe **one** reason why some paper and card products are deliberately not recycled by manufacturers. (3)
8 Explain how a manufacturer would use trend forecasting when designing birthday cards. (3)
9 A school quiz team wants their school logo printing onto some T-shirts. Describe how heat transfer paper could be used to produce the T-shirts. (4)
10 Paper used for many products comes from 'managed forests'. What is meant by the term 'managed forest'? (2)

# 4 Polymers

## 4.1 Design contexts

Polymers have only been around for the last 100 years. Over this time, many different polymers have been developed, each with their own unique properties. Designers and manufacturers need to understand polymers so that they can make an informed decision as to when and where they should be used.

## 4.2 Sources, origins, physical and working properties, and social and ecological footprint

### 4.2.1 Thermoforming polymers (including 4.2.4 Physical characteristics and 4.2.5 Working properties)

REVISED

Thermoforming polymers differ from thermosetting polymers in that they can be heated and moulded into shape many times.

Table 4.1 Thermoforming polymers

| Thermoforming polymer | Properties | Common uses |
|---|---|---|
| Acrylic (PMMA) | Hard, excellent optical quality, good resistance to weathering, scratches easily | Car light units, bathtubs, shop signage and displays |
| High-impact polystyrene (HIPS) | Tough, hard and rigid, good impact resistance, lightweight | Children's toys, yoghurt pots, refrigerator liners |
| Biodegradable polymer, or Biopol® | Lightweight, good electrical insulator, biodegradable | Disposable cups, razors, cutlery and packaging, surgical stitches and pins |
| High-density polystyrene (PS) | Lightweight, food safe, heat resistant | Food packaging, coffee cups, CD/DVD cases |
| Expanded polystyrene (EPS) | Lightweight, easy to mould, good impact resistance | Product packaging, disposable cups and plates |
| Extruded polystyrene foam (XPS), or Styrofoam™ | Lightweight, easy to work, good thermal insulation | Thermal insulation barrier in the construction industry, modelling material |
| Polyvinyl chloride (PVC) | Hard and tough, good chemical and weather resistance, low cost, can be rigid or flexible | Pipes, guttering, window frames |

Table 4.1 Thermoforming polymers (continued)

| Thermoforming polymer | Properties | Common uses |
|---|---|---|
| Acrylonitrile butadiene styrene (ABS) | High impact strength, lightweight, hard, durable | Safety helmets, safety glasses, mobile phone cases, car indicator lenses |
| Polyethylene terephthalate (PET) | Lightweight, strong, food safe | Bottles, food packaging |
| Urethane/polyurethane | High load resistance, high wear resistance, flexible | Bags, varnish, wheels, furniture foam |
| Fluoroelastomer | Heat resistant, chemical resistant, solvent resistant | Heat shrink tubing, car hoses, chemical-resistant gloves, gaskets |

# 4.2.2 Thermosetting polymers (including 4.2.4 Physical characteristics and 4.2.5 Working properties)

REVISED

Thermosetting polymers differ from thermoforming polymers in that once they have been formed, they cannot be reheated and reformed.

Table 4.2 Thermosetting polymers

| Thermosetting polymer | Properties | Common uses |
|---|---|---|
| Polyester resin | Good electrical insulator, good heat insulator, good chemical and wear resistance | Casting, boat hulls (when used as glass reinforced plastic, GRP), sports car bodies (when used as carbon fibre reinforced polymer, CFRP) |
| Urea formaldehyde (UF) | Stiff and hard, heat resistant, good electrical insulator | White electrical fittings, toilet seats, adhesive used in medium-density fibreboard (MDF) |

# 4.2.3 Sources and origins

REVISED

- Polymers are sourced from crude oil, a fossil fuel that is found deep underground.
- Crude oil is found all over the world but the largest reserves have been found in Russia, the United Arab Emirates and Saudi Arabia.
- The ground is drilled and the crude oil is pumped to the surface; it is then transported to a refinery where it is converted into polymers.

# 4.2.6 Social footprint

REVISED

## Trend forecasting

- Trend forecasters aim to predict the needs of polymer-based products in future years.
- They carry out research by surveying target markets, interviewing focus groups and investigating developments into new polymers.
- The development of biopolymers and 3D printing are areas that will change the way designers use polymers to produce unique, environmentally friendly shapes.

## Impact of extraction and production on environment and wildlife

- Crude oil is found underground and quite often far out at sea.
- Drilling for crude oil provides employment and can lead to the development of new communities.
- Drilling for oil on land can require forested areas to be cleared, destroying the habitats of many birds and wildlife. **Deforestation** is a major contributor to global warming.
- Drilling at sea can lead to oil spills, which are harmful to sea and bird life.
- Transporting oil around the world by ships can lead to spillages, which can wash up on beaches.
- Processing crude oil into polymers can release toxic gases that contribute to global warming.

## Recycling and disposal

- Most thermoforming polymers can be recycled.
- Polymers are identifiable by a recycling symbol that is usually moulded into the surface.
- Thermosetting polymers are difficult to recycle and generally end up in landfill where they take many years to decompose.
- Biopolymers will decompose if sent to landfill and do not harm the environment.

## 4.2.7 Ecological footprint  `REVISED` ☐

### Sustainability

- Most polymers come from crude oil, which is a finite resource that may run out in 50 years unless new reserves are discovered or extraction methods improve.
- Most thermoforming polymers are recyclable.
- Biopolymers come from plant-based materials, such as corn starch and sugar beet. They are infinitely renewable.
- Thermosetting polymers are not recyclable and are therefore not considered to be sustainable.

### Oil exploration and extraction

- To find oil reserves, geologists survey areas of land and the seabed by drilling test holes.
- Oil is extracted by pumping it to the surface.

### Processing

- Crude oil is processed at an oil refinery.
- There are three processes involved in refining crude oil to produce polymers: fractional distillation, cracking and polymerisation.

> **Deforestation**: the large-scale felling (cutting down) of trees that are not replanted.

PETE

HDPE

V

LDPE

PP

PS

Other

**Figure 4.1 Polymer recycling symbols**

### Transportation

- Oil can be transported to a refinery by ship or by pipeline.
- Polymers are then transported to factories in pellet form, usually by road or rail.

### Wastage

- Most waste from thermoforming polymer production can be recycled.
- Waste from thermosetting polymers will end up in landfill.

### Pollution

- The extraction, processing and transportation of polymers creates a significant amount of pollution, mainly in the form of toxic gases, such as carbon dioxide, being released into the atmosphere.
- Oil spillages can have a devastating effect on the sea, wildlife and environment.
- Non-recyclable polymers go to landfill and can harm plants and wildlife.
- Plastic waste frequently finds its way into the sea where it breaks down into microplastics (particles smaller than 5 mm) and can enter the food chain.

---

## Now test yourself                                              TESTED ☐

1  Give **three** properties of polyvinyl chloride (PVC) that make it suitable for a window frame.  (3)
2  Explain the difference between a thermoforming polymer and a thermosetting polymer.  (2)
3  Name a suitable polymer for an illuminated shop sign.  (1)
4  Explain why urea formaldehyde (UF) is an unsustainable polymer.  (4)
5  Describe the ecological footprint left by sourcing and processing crude oil from the sea.  (6)

---

# 4.3 Selection of polymers

## 4.3.1 Aesthetic factors                                        REVISED ☐

- **Form**: polymers can be manipulated in a variety of ways to change their form and shape, from thin, flat sheets of acrylic to complex, intricate 3D-printed products.
- **Colour**: polymers can be transparent, translucent or opaque. With the addition of a pigment, any colour of polymer can be achieved.
- Texture: a texture can be moulded into the surface, for example to add feel and provide grip on a handle.

> **Form**: the shape, size and proportion of a product.

**Figure 4.2 A cordless drill with a textured handle and mould grips**

# 4.3.2 Environmental factors

REVISED

## Sustainability

- Crude oil is a finite resource and it will eventually run out.
- If we recycle thermoforming polymers, then we increase their sustainability.
- Biopolymers are renewable, making them sustainable.
- Thermosetting polymers are not recyclable, and we cannot sustain their use indefinitely.

## Pollution

- Oil spillages pollute the sea and land, and are harmful to wildlife.
- The processing of crude oil pollutes the atmosphere, contributing to global warming.
- Polymers that end up in landfill pollute the land.
- Polymers can pollute the sea and end up in the food chain.

## Biodegradable polymers

- Biopolymers do less harm to the environment when they are being processed than traditional polymers.
- Biopolymers are biodegradable.

# 4.3.3 Availability factors

REVISED

## Use of stock materials

- Polymers are available in many stock forms such as powders, granules, sheets, rods, bars and tubing.
- Using stock forms of a material in your design can significantly reduce the cost of a product.

## Use of specialist materials

- **Thermochromic pigments** change the colour of a polymer when they are heated.
- **Phosphorescent pigments** glow in the dark.
- GRP and CFRP produce very strong and lightweight structures.

> **Thermochromic pigment**: a pigment that changes colour with heat.
>
> **Phosphorescent pigment**: a pigment that glows in the dark.

## Effect of global oil supply

The global supply of oil is dependent on several influential factors: supply and demand, politics, conflicts between warring nations and trade wars.

# 4.3.4 Cost factors

REVISED

## Quality of material

- The quality and type of material used has a direct impact on the cost of the final product.

- For example, if a Formula 1 racing car was made from GRP the cost would be significantly cheaper. However, as money is not a prime concern of an F1 team, they can use CFRP, which is more expensive but has better qualities.

## Manufacturing processes

- Some processes, such as moulding using GRP, are costly as they are very labour intensive.
- Highly mechanised processes, such as injection moulding, have a high initial set-up cost. But if the product is needed in high quantities then the initial cost of each product is relatively inexpensive.

## Treatments

Specialist treatments and additives – such as adding fillers, ultraviolet (UV) stabilisers, foaming agents, plasticisers, fire-proofing agents and pigments – will add to the cost of the polymer.

## Commodity price

The cost of oil is measured in US dollars per barrel and is set by the Organisation of the Petroleum Exporting Countries (OPEC).

## 4.3.5 Social factors

REVISED

- Different social groups can influence the style and design of polymer products. A TV remote control can be themed to a particular football team, and game controllers can be customised to a particular type of game.
- Trends and fashions change over the years. In 1942 Earl Tupper developed an airtight polythene container, starting a brand of products called Tupperware, which changed the way we take our packed lunches to school/work.
- A product's popularity can have a direct effect on the type of polymer used. PET is used in the production of plastic water bottles as it is food safe, lightweight, fully recyclable and can be produced in high volumes at little cost.

## 4.3.6 Cultural and ethical factors

REVISED

### Avoiding offence

- Polymer packaging is being washed up on remote islands in the Indian Ocean. This is not only harming wildlife but also ruins the home of the people who live there.
- A cheap plastic native American headdress may seem fun to wear at a fancy-dress party but it is considered to be culturally offensive.

### Suitability for intended market

As polymers are relatively easy to mould into unique shapes it makes them an ideal material to use when special adaptations are needed. CFRP can be used to make prosthetic limbs, such as blades.

**Figure 4.3 A prosthetic leg**

## Use of colour and language

The majority of smart phone cases are either black or white as these colours are the most requested ones and therefore help the phones to sell to any culture in any country.

## Consumer society

There is an ever-growing demand by society for new and improved products. Polymer products can fill this demand as it is relatively quick and easy to produce a new design and have it manufactured in quantity.

## Effects of mass production

● Polymer products are perfectly suited to mass production. This method significantly reduces the cost of manufacturing, making products affordable to more people.

● However, mass production means that fewer workers are needed, which could lead to unemployment.

## Built-in product obsolescence

● Most polymer products are built with a limited life. This ensures that customers will frequently return to replace old, obsolete products.

● For example, your ballpoint pen contains a certain amount of ink and, when it runs out, you will need to buy another one.

> **Exam tip**
>
> When asked to explain the advantages of using a particular polymer with a certain product, make sure you justify your reasons.

> **Typical mistake**
>
> When candidates are asked to give both advantages and disadvantages, they will simply reverse an advantage to make it into a disadvantage, and this will often not be given a mark.

### Now test yourself
TESTED ☐

1 Give the definitions of the terms 'transparent', 'translucent' and 'opaque'. (3)
2 Explain why texture is applied to handles. (1)
3 Give **three** factors that can affect the global supply of crude oil. (3)
4 Describe the effects of mass production on polymer-based products. (3)
5 Discuss the sustainability of biopolymers. (4)

# 4.4 Forces and stresses

## 4.4.1 Forces and stresses
REVISED ☐

**Table 4.3 Forces and stress acting on polymers**

| Type of force | Definition | Example |
|---|---|---|
| Compression | A force that is pushing down on an object | Underground polymer piping used in water and sewage systems |
| Tension | A force that is pulling an object apart | Rope, shopping bags, parcel tape |
| Shear | A force that acts in opposite directions | Bank notes, camping chair |
| **Flexibility** | The ability to bend and deform without breaking | Food bags, food wrapping |

> **Flexibility**: the suppleness and bendability of a polymer.

> **Typical mistake**
>
> Candidates often confuse the terms 'compression', 'tension' and 'shear'. Make sure you know the difference.

## 4.4.2 Reinforcement/stiffening techniques

- **Frame structures**: GRP can withstand compressive, tensional and shear forces. It is also flexible, making it ideal for use as a frame structure for a pop-up tent.
- **Triangulation**: a triangle makes a very strong, rigid structure. When force is applied to the top of the triangle, two sides act in compression while the third will be in tension, but the structure will not change.
- **Fabrication, assembly and construction processes**: webs, bends, ridges and creases can be moulded into thin-walled polymer shapes to help strengthen them.
- **Additives**: polyester resins can be strengthened significantly by adding glass fibre to form GRP or carbon fibre to form CFRP.

**Triangulation**: the strengthening of a structure using triangles.

**Fabrication**: the cutting, shaping and joining of components.

**Additives**: materials that are added to a polymer to improve its properties.

### Now test yourself

**TESTED**

1  What is the force being applied to a nylon climbing rope? (1)
2  Explain how a thin-walled moulding can be strengthened. (3)
3  What is meant by the term 'triangulation'? (2)

# 4.5 Stock forms, types and sizes

### Exam tip

Make sure that you can show, using both notes and sketches, how polymers can be strengthened with the use of webs, bends, ridges and creases during the moulding stage.

## 4.5.1 Stock forms/types and 4.5.2 Sizes

**Table 4.4** Stock forms of polymers

| Stock form | Description | Use |
|---|---|---|
| Bar | A solid, round or square, cross-sectional polymer<br><br>Usually sold in lengths of 1–4 m | General purpose fabrication work |
| Sheet | A 1000 mm × 600 mm sheet of polymer<br><br>Usually available in varying thickness of up to 10 mm | Vacuum forming<br><br>Illuminated shop signage |
| Pipe and tube | Polymer pipe and tubing with a hollow cross-section<br><br>Usually sold in lengths of 1–4 m with varying wall thicknesses | Water pipes, structural framework |
| **Mouldings** | Specialist moulded components | Guttering, door frames, window frames, shower trim |
| **Resin** | A liquid polymer that becomes solid when mixed with a hardener<br><br>Sold by weight | Cast artwork<br><br>Boat hulls and sports car bodies when mixed with glass fibre (GRP) |

**Mouldings**: specially shaped polymer components.

**Resin**: liquid polymer.

**Table 4.4** Stock forms of polymers (continued)

| Stock form | Description | Use |
|---|---|---|
| **Granules/** powder | Fine particles of polymer<br><br>Can be injection moulded or extruded<br><br>Can be sprayed on to a hot surface where it melts and acts as a polymer coating<br><br>Sold by weight | Injection moulded/ extruded polymer products<br><br>Powder coating and plastic dip |
| **Film** | A very thin sheet of polymer usually obtained in a roll<br><br>Sold in varying widths and lengths | Food wrapping, vinyl stickers |

**Granules**: grain-sized polymers used in injection moulding and extrusion.

**Film**: very thin polymer sheets.

### Exam tip

Make sure that you can name an application of each different stock form of polymer.

### Typical mistake

When asked to calculate the cross-sectional area of tubing, candidates often simply give the cross-sectional area of the outside diameter.

## Now test yourself

TESTED ☐

1  What is the cross-sectional area of a polymer bar with a diameter of 20 mm? (1)
2  Name the stock form of a polymer used to make food wrapping. (1)
3  Give an example of a polymer resin. (1)

# 4.6 Manufacturing to different scales of production

## 4.6.1 Processes

REVISED ☐

Polymers are suited to industrial processes where products can be made in very large quantities.

### Blow moulding

**Blow moulding** is used to produce hollow polymer products such as water bottles. The blow moulding process is:

1  A heated parison (tube) is lowered between the two halves of a mould.
2  The mould is closed and air is blown into the mould, forcing the hot parison to the outsides of the mould.
3  The mould is cooled, opened and the finished product is removed.

**Blow moulding**: a method of moulding by blowing a heated polymer into shape.

**Press moulding**: a method of shaping a polymer by heating it and pressing it into a mould.

### Press moulding

**Press moulding** is used to produce solid polymer products such as electrical plugs and sockets. The press moulding process is:

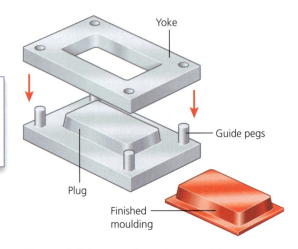

**Figure 4.4 A press forming mould**

Answers, glossary and quick quizzes at www.hoddereducation.co.uk/myrevisionnotes

1  A preform (pellet) of polymer is placed between two halves of a mould.

2  The mould is heated and the upper half of the mould is lowered, compressing the preform into the mould cavity.

3  The mould is cooled, opened and the finished product is removed.

## Extrusion

**Extrusion** is used to produce intricate cross-sections of polymer such as guttering.

> **Extrusion:** a long length of polymer with a consistent cross-section.

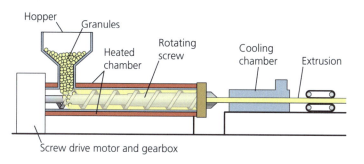

Figure 4.5 **The extrusion process**

## Injection moulding

**Injection moulding** produces solid polymer shapes such as school chairs. The injection moulding process is:

1  Polymer pellets are fed from a hopper into a heating chamber.

2  The pellets are transported along the chamber via an Archimedean screw.

3  The pellets become molten and are then injected into a mould.

4  The mould is cooled, and the moulded product is removed.

> **Injection moulding:** an industrial method of heating a polymer and injecting it into a mould.

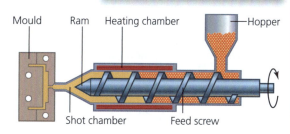

Figure 4.6 **An injection moulding machine**

## Polymer welding

- Thermoforming polymers can be welded together in a variety of ways to form a permanent joint.
- Each method has its own way to heat the polymer surfaces so that they will fuse (melt) together:
  - **laser polymer welding** uses a laser to heat the polymer
  - **ultrasonic welding** uses ultrasonic vibrations to generate heat
  - **hot gas welding** uses a hot flame to melt the surface of the polymer.

## Line bending

Line bending produces a straight-line bend in a thermoforming polymer. The line bending process is:

1  A thermoforming polymer sheet is marked out with a bending line.

2  The line is placed directly above the hot wire that is located within a strip heater.

3  When the sheet becomes soft, it can be bent to any desired angle.

4  The sheet must now be held until it sets.

Figure 4.7 **A strip heater**

## 4.6.2 Scales of production

**Table 4.5 Scales of production for polymers**

| Scale of production | Advantages | Disadvantages | Examples |
|---|---|---|---|
| One-off | A high-quality, unique product is produced | Usually involves a specialised process requiring a highly skilled workforce<br><br>Expensive | A shop frontage sign for a specialist outlet<br><br>A piece of art cast in resin |
| Batch | A limited number of identical products are produced<br><br>The cost of the product is reduced as materials can be bought in bulk and processing times are reduced | There is an initial set-up cost involving the purchase of machinery and manufacturing aids | Themed or seasonal products such as Christmas decorations |
| Mass | Large quantities of identical products are produced<br><br>The cost is further reduced as computer-aided manufacture (CAM) can be utilised | High initial set-up cost as there is intensive use of machines<br><br>Products lose their individuality | Car parts, such as indicator lenses, dashboards and switches |
| Continuous | The unit cost of a product is significantly reduced as materials can be purchased in vast quantities<br><br>A very high demand can be satisfied | Initial set-up costs are very high, often requiring a dedicated factory to be built<br><br>There is likely to be extensive use of robotics and CAM | Water bottles |

## 4.6.3 Techniques for quantity production

### Marking out

- A spirit-based pen and a plastic ruler should be used if you are marking out directly on to the surface of a polymer. This will prevent scratching on the surface.
- A laser cutter can engrave marking out lines on to the surface of most polymers with great accuracy.

### Jigs

- Jigs speed up production and increase accuracy by removing some of the marking out processes.
- A bending jig will ensure that a bend is accurately and consistently produced when forming acrylic (PMMA) on a strip heater.

### Templates

- Templates are generally made from paper and stuck on to the surface of a polymer. They can be made from more resilient materials and drawn around.

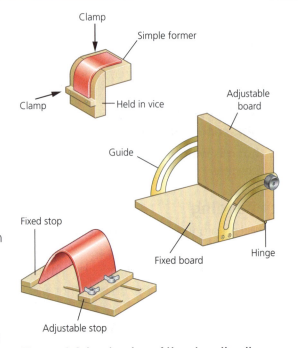

**Figure 4.8 A selection of line-bending jigs**

- Templates reduce the need for time-consuming marking out and are particularly useful when several identical products need to be made.

## Patterns

- When casting in a resin, a pattern is initially produced from an inexpensive material such as MDF.
- The pattern is then used to produce identical moulds.

## Moulds

- Moulds are used when a batch of identical-shaped products are required.
- Moulds can be used when resin casting, vacuum forming or when producing a GRP or CFRP shell.

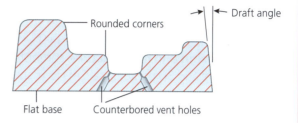

**Figure 4.9 A cross-section through a mould used for vacuum forming**

## Computer-aided manufacture

- Laser cutters and 3D printers are specialist machines that can be used to cut and form polymer-based materials.
- They take their information from computer-aided design (CAD) and work with accuracy, consistency and speed.
- CNC machinery, such as lathes and milling machines, are traditionally used with metals but can perform the same operations on polymers.

## Quality control

- Quality control checks should be carried out at regular intervals during the manufacturing process.
- This can involve removing products from a production line and checking them for dimensional accuracy, or by checking the machine to see if it is running correctly.

## Working within tolerances

- If a component is to interact with another component, it must be manufactured within a certain tolerance or it may be too big or too small to function.
- A tolerance is given as a size plus or minus an acceptable limit.

## Efficient cutting to minimise waste

The efficient arrangement of multiple CAD drawings can significantly minimise waste. This is known as **tessellation**.

> **Tessellation**: the arrangement of components to minimise waste.

### Now test yourself

TESTED ☐

1. Use notes and sketches to explain the process of extrusion. (6)
2. Name the equipment you would need to produce a bend in acrylic (PMMA). (1)
3. Explain the advantages of continuously producing polymer products. (4)
4. Give an example of where you would use a mould when producing polymer products. (1)

# 4.7 Specialist techniques, tools, equipment and processes

## 4.7.1 Tools and equipment

- **Hand tools and machinery**: many traditional woodworking and metalwork hand tools and machines can be used to shape polymers. For more information see Section 2.7.1 Tools and equipment on page 64 and Section 7.7.1 Tools and equipment on page 146.
- **Digital design and manufacture**: computer-aided machines such as laser cutters and 3D printers use CAD to manufacture polymer products quickly, accurately and consistently.

## 4.7.2 Shaping

### Laser cutting and engraving

- Laser cutters use CAD drawings to cut and engrave polymers such as acrylic (PMMA).
- It is quick, accurate and consistent, and requires little human intervention.

### Cutting

- Most woodworking and metalworking saws will cut polymers. For more information on these tools see Section 2.7.1 Tools and equipment on page 64 and Section 7.7.1 Tools and equipment on page 146.
- Thin polymer sheets can be cut using a craft knife.
- Expanded polystyrene (EPS) can be cut using a hot wire cutter.

### Filing

- Metalworking files work well on polymers. For more information on files and filing techniques see Section 2.7.2 Shaping on page 64.
- When holding polymers in a vice or in a clamp, they must be protected to prevent them from scratching.

### Bending

- Polymers are usually bent using a strip heater.
- Polymers can also be bent by softening them in an oven and then bending them over a former.

### Abrading

- Emery cloth can be used to smooth the edges of polymer sheet.
- Very fine grades of wet and dry paper will give a smoother finish.

### Vacuum forming

- **Vacuum forming** polymers produces a thin-walled 3D shape from a pre-manufactured mould.

> **Abrading**: smoothing the surface of a material with abrasive papers.
>
> **Vacuum forming**: a method of shaping polymers by heating and sucking around a mould.

- Yoghurt pots, margarine tubs and the insides of fridges are all examples of vacuum-formed products. For more information on vacuum-forming moulds see Section 4.6.3 Techniques for quantity production on page 99.
- The vacuum-forming process:
  1 Once the mould is ready, it is placed on the platen (table) of the vacuum former and lowered into the machine.
  2 A sheet of high-impact polystyrene (HIPS) is then clamped over the top of the machine and heat is applied.
  3 After a short time, the HIPS sheet will become soft. Care should be taken not to overheat the HIPS sheet as it will not form properly and webbing may occur.
  4 The mould is then raised up into the hot HIPS sheet and the air is immediately sucked out by turning on the vacuum pump.
  5 Once formed, the sheet should be allowed to cool then removed from the vacuum former and trimmed.
  6 Deeper moulds may require the soft HIPS sheet to be blown into a dome before the mould is raised. This gives an even thickness of material around the taller mould.

## Deforming and reforming

- Polymers are ideally suited to shaping by deforming. Once heated they can be bent, formed around a mould, vacuum formed, blow moulded or moulded by compression.
- A number of industrial processes can reform polymers; these include injection moulding, extrusion, casting and 3D printing.

# 4.7.3 Fabricating/constructing/assembling

REVISED

## Tapping and threading

Polymers can be **tapped** and **threaded** using the same tools and methods that you would use on metals. For more information see Section 2.7.4 Assembling on page 67.

## Fastening

- Traditional nuts, bolts and washers can be used with polymers.
- Polymer-specific bolts have a larger head to disperse the pressure and to avoid damaging the surface.
- For more information see Section 2.7.4 Assembling on page 68.
- Panel trim fittings are polymer-specific **fasteners** that are used extensively in the manufacture of cars.

## Adhesives

- **Adhesives** can be used to provide a permanent joint.
- You must ensure that the surfaces to be glued are clean, dry and free from dust, oil and dirt.
- When using adhesives, avoid contact with the skin and eyes and avoid inhalation.

> **Tapping**: a method of producing internal screwthreads.
>
> **Threading**: a method of producing external screwthreads.
>
> **Fasteners**: polymer clips that are used to hold plastic panels together.
>
> **Adhesives**: glues.

**Table 4.6** Polymer adhesives

| Adhesive | Use | Application | Advantages | Disadvantages |
|---|---|---|---|---|
| Contact adhesive | Gluing large surface areas | Apply a thin coat to both surfaces<br><br>Allow time to dry<br><br>Bring both surfaces together | A join is created on contact<br><br>Can be used to glue dissimilar materials together | Can only be used effectively on large surface areas<br><br>Highly flammable when in liquid form |
| Epoxy resin | For gluing polymers to polymers or any other material | Mix equal amounts of adhesive and hardener<br><br>Apply to one surface<br><br>Clamp together until set | A strong joint is created<br><br>Can be used to glue dissimilar materials together | Expensive<br><br>Must be mixed carefully |
| Tensol® cement | For gluing polymers to each other | Apply a thin coat to one surface<br><br>Clamp together until set | Dries clear<br><br>Fuses the surfaces together | Expensive<br><br>Highly flammable when in liquid form |
| Liquid cement (superglue) | For gluing polymers to polymers or any other material | Apply a thin coat to one surface<br><br>Hold together until set | Very quick<br><br>Dries clear<br><br>Fuses the surfaces together | Expensive |

**Typical mistake**

When asked to describe a method of applying an adhesive, candidates often leave out vital information regarding the preparation of the surfaces.

**Exam tip**

Make sure that you can produce a detailed explanation, using both notes and sketches, of how to apply each different type of adhesive.

## Wastage

- Wastage involves the removal of polymer material to produce a shape.
- All the cutting, shaping and abrading tools covered in this chapter deal with wastage.

## Addition

- Addition involves adding material to produce a polymer product.
- All the fabrication methods covered in this chapter relate to addition.

## Now test yourself

TESTED ☐

1. Use notes and sketches to describe the process of vacuum forming. (6)
2. How do nuts, bolts and washers that are specific to polymers differ from traditional nuts, bolts and washers? (1)
3. Give an advantage of using Tensol® cement. (1)
4. Use notes and sketches to describe the process of gluing using contact adhesive. (4)
5. What are the health and safety issues concerned with using liquid solvent cement? (3)

# 4.8 Surface treatments and finishes

- Most polymers are supplied self-coloured with a high-quality finish.
- They are both waterproof and chemical resistant, and therefore need less finishing than timber and metal.

## 4.8.1 Surface finishes and treatments  REVISED

### Polishing

- The edges of acrylic (PMMA) can be **polished** by firstly draw filing with a fine file and then polishing with a mildly abrasive polish.
- Polymer surfaces can be buffed to a high shine using a buffing machine.

### Textured surfaces

Textures can be formed on to the surface of a polymer at the moulding stage. This is particularly useful on tool handles to provide them with grip.

### Laser engraving

Textures and images can be engraved on to the surface of polymers such as acrylic (PMMA).

### Vinyl stickers

A vinyl cutter can be used to produce vinyl stickers. Vinyl cutters use CAD drawings to direct a knife through self-adhesive vinyl.

### GRP pigments

**Coloured pigments** can be added to polyester resin to colour products made from GRP.

> **Polishing**: rubbing the surface of a material to achieve a shiny finish.

> **Typical mistake**
> When asked a question relating to finishing polymers, candidates often forget that finishes can also provide texture.

> **Exam tip**
> A laser cutter can be used to engrave as well as cut. High-ability candidates can quote the different power settings for each process.

> **Laser engraving**: using a laser to etch the surface of a polymer.
>
> **Coloured pigments**: a dye that can be added to resins to change their colour.

## Now test yourself  TESTED

1 Use notes and sketches to describe how to finish the edges of acrylic. (4)
2 Name the equipment needed to produce a vinyl sticker. (1)
3 What is added to polyester resin to change its solvent colour? (1)

## Exam practice

1 Give **two** properties of expanded polystyrene (EPS) that make it suitable for use as packaging. (2)
2 Give **two** properties of acrylonitrile butadiene styrene (ABS) that make it suitable for the casing of a mobile phone. (2)
3 Give the source material of biodegradable polymers. (1)
4 Explain what is meant by the term 'planned obsolescence'. (3)
5 Use notes and sketches to describe the process of blow moulding. (6)
6 Use notes and sketches to describe the process of line bending acrylic (PMMA). (4)
7 Describe the advantages of using a jig when bending acrylic (PMMA). (3)
8 Explain the advantages to the manufacturer of using stock forms of polymers. (3)
9 Discuss the sustainability issues concerning the use of thermosetting polymers. (6)
10 Give the material source of polymers. (1)

# 5 Systems

## 5.1 Design contexts

The **design context** is the wider setting in which a design solution for a product or system will sit. It is the starting point from which a specific design problem is identified, outlined and investigated. This leads to the identification of a need for a product or system that could solve the problem.

Examples of design contexts could include:

- dealing with the impacts of natural disasters and/or climate change
- supporting people with learning disabilities
- encouraging people to lead more active lifestyles
- ensuring safety in the workplace.

### End users

It is important to understand the needs of the intended **end user**. One way is through **market research**, such as interviews, observation and focus groups. It is also helpful to produce a **user profile**.

### Considerations when designing and modifying a product

The context should be considered carefully at all stages to ensure that the solution is fit for purpose. For example, for a design context 'encouraging people to live more sustainable lives' considerations could include:

- Are the selected materials and components easy to recycle?
- Can the components be removed easily for reuse in a different product or system?
- Could a more energy-efficient manufacturing process be used?

> **Design context**: the wider setting in which a design solution for a product will sit.
>
> **End user**: the person who will purchase or use the finished product.
>
> **Market research**: the process of gathering information about the wants and needs of potential users.
>
> **User profile**: an overview of the intended user containing information such as their age, occupation and general interests.

> **Exam tip**
>
> Ensure that you understand the importance of the design context to the designing and making process.

> **Typical mistake**
>
> Confusing the design context with the design brief. The context gives the wider setting, whereas the brief gives an overview of the exact problem that is to be solved.

> **Now test yourself**    TESTED ☐
>
> 1 Define the term 'design context'. (1)
> 2 Give **two** examples of design contexts. (2)
> 3 Explain why the design context must be considered at all stages of the design and manufacturing process. (2)

## 5.2 Sources, origins, physical and working properties, and social and ecological footprint

### 5.2.1 Sensors    REVISED ☐

- **Sensors** take signals from the environment around them and turn them into electronic signals.

> **Sensor**: a component that detects changes in the environment around it.

- For information about LDRs and thermistors, see Section 1.6 Electronic systems on page 23.
- Moisture sensors are used to detect the presence of water, for example in soil. A voltage is created by the sensor that is proportional to the moisture content detected.
- Piezoelectric sensors are used to measure changes in pressure, force, strain and acceleration. This is achieved by turning these into an electrical charge using the piezoelectric effect.

## 5.2.2 Control devices and components

REVISED

For more information about resistors, transistors and switches, see Section 1.6 Electronic systems on page 23.

### Switches

- Switches are used to 'make' or 'break' a circuit. The contacts inside must be joined together for the current to flow between them.
- With rocker switches, the contacts are joined when the switch is pressed into the 'on' position. Current flows until the switch is pressed back into the 'off' position.
- With push to make (PTM) switches, the contacts are joined when the switch is pressed.
- Micro switches are extremely sensitive switches that respond to small movements.
- Reed switches are magnetic switches. The contacts close in the presence of magnetism.

**Figure 5.1 A rocker switch**

### Resistors

There are two main types of resistor:

- **Fixed resistors** have a set value that does not change.
- **Variable resistors** have resistance that can be adjusted using either a small screwdriver hole or a spindle with an operating knob.

**Figure 5.2 A micro switch**

### Control components

- **Bipolar transistors** are used as electronic switches and current amplifiers. There are two main types: negative-positive-negative (NPN) and positive-negative-positive (PNP).
- A **microprocessor** is a programmable integrated circuit that controls the processing functions of a computer system. It acts like the brain of the system.
- A **microcontroller** is a small computer on an integrated circuit. It has pins for the connection of different input and output devices. It can be programmed to complete tasks such as reading analogue sensors, timing and counting. A commonly used example is a peripheral interface controller (PIC).
- **Relays** are used to switch a large load current from a much smaller control current. They allow one circuit to switch a second, completely separate, circuit.

**Microcontroller**: a small computer on an integrated circuit that can be programmed to perform different functions.

**Figure 5.3 A PIC microcontroller**

## 5.2.3 Outputs

REVISED

- Outputs take electronic signals and turn them back into 'real world' signals.
- For information about buzzers and LEDs, see Section 1.6.3 Outputs on page 23.
- **Loudspeakers** convert an electrical audio signal into sound. They require a driver circuit to provide the signal.
- **Motors** convert current into rotary motion using electromagnetic induction.

> **Loudspeaker**: a component that converts an electrical audio signal into sound.
>
> **Motor**: a component that converts current into rotary motion.

## 5.2.4 Sources and origins

REVISED

**Table 5.1 Origins of materials and components used in systems**

| Material | Countries of origin/manufacture |
|---|---|
| Polymers from crude oil, e.g. acrylic, high-impact polystyrene (HIPS), acrylonitrile butadiene styrene (ABS) | Russia, Saudi Arabia, USA |
| Silicon | China, Russia, USA |
| Gold | China, Australia, Russia |
| Copper | Chile, China, Peru |
| Lithium | Australia, Chile, Argentina |
| Aluminium | China, Russia, Canada |
| Rare earth elements (REEs) | China, Australia, USA |
| Nickel | Philippines, Indonesia, Russia, Canada, Australia |

**Figure 5.4 A loudspeaker**

## 5.2.5 Physical characteristics

REVISED

- Components often have ratings, values and/or **tolerances** assigned to them.
- For example, each fixed resistor has a value in ohms. The higher this value, the more resistance it has.
- With a resistor, the tolerance is often between 5 per cent and 20 per cent, but it can be lower.
- The resistor colour code is described in Section 5.8.1 Surface finishes and treatments on page 115.
- Ratings give the voltage that a component or system is designed to work at, and the current that will be used at that voltage.
- When selecting materials to construct cases for systems, their physical and working properties, sustainability and potential manufacturing processes must be considered.

> **Tolerance**: the allowable amount of variation of a stated measurement.

> **Exam tip**
>
> Make sure you know the different materials used in electronic systems and where they are sourced from.

## 5.2.6 Working properties

- Thermal conductivity is the ability of a material to transmit heat.
- Electrical conductivity is the ability of a material to conduct electrical current.
- **Insulators** are materials that do not conduct heat or electricity well, whereas conductors do.
- When selecting materials for system cases, their durability, hardness, toughness and elasticity must be considered.

> **Insulator**: a material that does not conduct heat or electricity easily.

## 5.2.7 Social footprint

A **social footprint** is the impact of an activity on society, for example:

- using scarce and/or hazardous elements in systems, such as cobalt, tantalum and lithium
- the effects of using modern communication systems such as mobile phones, computers, games consoles and social media networks.

> **Social footprint**: a measure of the impact of an activity on society.

## 5.2.8 Ecological footprint

An **ecological footprint** is the impact of an activity on the local ecosystem, for example:

- the damage caused to habitats when elements are extracted or processed
- built-in obsolescence resulting in more landfill waste
- the effects of the use of products and systems, such as pollution
- the toxicity of metals and polymers, leading to poisoning of the environment when disposed of.

> **Ecological footprint**: a measure of the impact that human activity has on the environment.

### Now test yourself

TESTED

1  Name a sensor that could be used to measure the amount of water in soil. (1)
2  Describe what is meant by the tolerance of a resistor. (2)
3  Name **three** countries where silicon can be sourced from. (3)
4  Explain the difference between thermal and electrical conductivity. (2)
5  Explain **one** example of the impact of the use of communication systems on society. (2)

# 5.3 Selection of systems

## 5.3.1 Aesthetic factors

- **Aesthetics** includes the form (or shape) of a product, the colours used and its physical texture.
- Cases for systems should be aesthetically appealing to the end user.

> **Aesthetics**: how something looks and feels.

## 5.3.2 Environmental factors

- The Restriction of Hazardous Substances (RoHS) Directive prohibits the inclusion of certain hazardous substances in electronic products. This includes lead, mercury and cadmium.
- The Waste Electrical and Electronic Equipment (WEEE) Directive aims to reduce the amount of electronic waste that is sent to landfill by placing responsibilities on the producers of products.

### 5.3.3 Availability factors

- Materials and components come in **stock forms**. For example, microcontrollers commonly come in 8, 14, 16, 18, 20, 28 and 40 pin versions.
- Some materials are unusual or unique, or must be specially produced for a system.
- Some elements used to create components, such as cobalt, are scarce and difficult to obtain.

> **Stock forms**: the standard shapes and sizes that a material or component is available in.

### 5.3.4 Cost factors

- The higher quality a component is, the more it will cost. Similarly, components with tighter value tolerances are more expensive than those with greater variation.
- The comparative cost of different manufacturing processes must also be considered.

**Figure 5.5 Microcontrollers in 8, 16 and 28 pin forms**

### 5.3.5 Social factors

- Systems can be designed for the benefit of different social groups, for example mobile phones with large buttons for use by older adults. However, products can also have a negative impact on society.
- Other social factors include current trends and fashions, and the wider popularity of products.

### 5.3.6 Cultural and ethical factors

These are the beliefs, moral values and traditions that affect the design of a product and its suitability for the intended market. For example:

- ensuring the product does not cause offence to different religious groups
- ensuring the use of colour and language is appropriate.

The effects of the 'consumer society', mass production and built-in obsolescence must also be considered.

> **Ethical factors**: the beliefs, moral values and traditions that affect the design of a product.

> **Exam tip**
>
> Make sure you know the main factors that influence the selection of materials and components for systems, and understand how they influence it.

> **Now test yourself** TESTED
>
> 1 State **two** stock forms for microcontrollers. (2)
> 2 Explain what is meant by a 'scarce' element. (2)
> 3 Give **two** examples of cultural factors that affect the design of systems. (2)

# 5.4 Forces and stresses

## 5.4.1 Forces and stresses

- **Tension** is a pulling or stretching force. It causes objects to become elongated or can even pull them apart.

> **Tension force**: a pulling or stretching force.

- **Compression** is a pushing force. It attempts to reduce the size of objects. Compressive stresses can cause buckling of structures.
- **Torsion** is a twisting force. It occurs when one end of an object is twisted and the other end is held in place or twisted in the opposite direction.
- **Shear forces** push one part of an object in one direction and another part in the opposite direction.

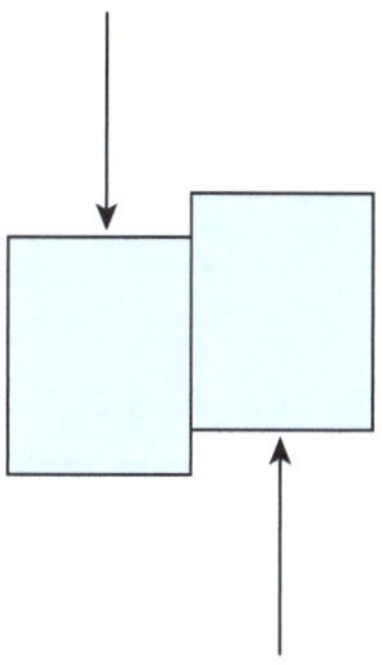

## 5.4.2 Reinforcement/stiffening techniques

REVISED

**Figure 5.6 Shear forces**

- **Composite materials** are created through the combination of two or more different types of material. They can result in unique combinations of properties that would not be possible to achieve when using individual materials.
- An example of a composite material is carbon fibre reinforced polymer (CFRP). This material has a high strength-to-weight ratio and high resistance to corrosion. It is increasingly being used in aircraft design.
- **Ribbing** is a method used to strengthen case structures. The ribs are placed in different formations to create patterns that stiffen the structure. It is a technique often used with plastic casings.

> **Compression force**: a pushing force.
>
> **Torsion force**: a twisting force.
>
> **Shear force**: a force acting in opposite directions.
>
> **Composite material**: a new material created by combining two or more different types of material.
>
> **Ribbing**: the use of ribs placed in different patterns to strengthen and stiffen a structure.

### Now test yourself

TESTED

1 Explain the difference between tension and compression. (2)
2 Give a specific example of torsion. (1)
3 Give an example of a composite material and state **two** of its properties. (3)

# 5.5 Stock forms, types and sizes

## 5.5.1 Stock forms/types

REVISED

### Resistor values and tolerances

- Resistors only come in certain values, within a specified tolerance. The E series is a system of preferred values that designers can select from.
- With the E12 series, there are 12 resistor values for each power, or multiple, of ten. Each resistor in this series has a tolerance of ± ten per cent. Because of this, it is usually fine for a designer to select the nearest resistor in the series to what they have calculated as necessary.

### Circuit construction types

- Permanent, soldered circuit boards can be constructed using either the **through-hole component method** or **surface mount technology** (SMT).
- With through-hole components the circuit board has both a plastic side and a side with the copper tracks and pads on it, which make up the circuit layout. Each pad has a hole drilled through it. Components are placed on top of the plastic side, with the legs going through the holes to the copper side. They are then soldered on to the pads to make the electrical connections.

> **Surface mount technology**: a method of circuit construction where components are placed on the surface of the circuit board.

- SMT is a more modern method that aims to reduce the size of circuit boards and components. The tracks are on both sides. The components are placed on the surface of the board and joined to it using solder paste or glue.

## 5.5.2 Sizes `REVISED` ☐

### Current, resistance and potential difference

- **Current** is the flow of electric charge through a conductive medium, such as a wire. The unit of measurement for current is the amp, or A.
- **Resistance** is the amount of opposition to the flow of current. The unit of measurement is the ohm, or Ω.
- **Potential difference** is the difference in charge between two points. It is often referred to as voltage. The unit of measurement is the volt, or V.
- **Ohm's Law** states that the current flowing through a resistor is directly proportional to the potential difference across the resistor. This is represented using the formula:

Potential difference (V) = current (I) × resistance (R)

- This can be shown as a triangle to help with recalling it, as shown in Figure 5.7.
- Ohm's Law can be used to calculate any one of the three values if the other two are known. For example, if a 330 Ω resistor has a current of 0.03 A flowing through it, the potential difference across it would be:

V = 0.03 × 330 = 9.9 V

### Resistors in series

The total value of resistors connected in series is calculated using the formula:

$$R_{total} = R_1 + R_2 + R_3 \text{ etc.}$$

**Figure 5.8 Resistors in series**

### Resistors in parallel

The total value of resistors connected in parallel is calculated using the formula:

$$\frac{1}{R_{total}} = \frac{1}{R_1} + \frac{1}{R_2} + \frac{1}{R_3} \text{ etc.}$$

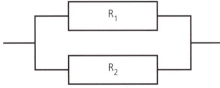

**Figure 5.9 Resistors in parallel**

---

**Exam tip**

Ensure that you understand and can explain the differences between tension, compression, torsion and shear forces.

---

**Current**: the flow of electric charge through a conductive medium.

**Resistance**: the amount of opposition to the flow of current.

**Potential difference**: the difference in charge between two points.

**Ohm's Law**: the relationship between potential difference, current and resistance, expressed as V = I × R.

---

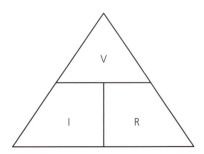

**Figure 5.7 Ohm's Law**

---

**Exam tip**

Make sure you can use Ohm's Law to calculate potential difference, current and resistance.

---

**Typical mistake**

When calculating resistor values in parallel, remember to perform the final $1/R_{total}$ part of the calculation, as this is often forgotten.

---

## Area and diameter

- Sometimes it is necessary to calculate the area of a component:

Area of rectangle = length × width

Area of a triangle = $\frac{1}{2}$ × base × height

Area of a circle = π × radius², or πr²

- If a straight line is drawn from one edge of a circle to another edge, through its centre, the length of this line is the diameter. The diameter is twice the length of the radius.

---

### Now test yourself
TESTED

1 Explain what is meant by through-hole circuit construction. (3)
2 State the units of measurement for current and potential difference. (2)
3 Define Ohm's Law. (1)
4 Three resistors of value 10 kΩ, 1.2 kΩ and 4.7 kΩ are connected in series. Calculate the total resistance of this arrangement. (2)

---

# 5.6 Manufacturing to different scales of production

## 5.6.1 Processes
REVISED

### Photo etching

**Photo etching** is a **printed circuit board** (PCB) production method:

1 A mask is placed on the underside of a piece of photo resist board. This is then put into a UV light box for a few minutes.
2 It is briefly placed into developer solution before being removed, washed and held in an etch tank.
3 The etching solution slowly removes the unwanted copper, leaving just the layout from the initial mask.

### PCB population

- Once a PCB has been produced it must be populated with components.
- If the through-hole method is being used, small holes are drilled in the centre of each copper pad.
- Once the component legs have been placed through the holes, they are soldered to the pads. The hot tip of the soldering iron is placed where the connection is to be made, heating it for a few seconds. Solder is then applied.
- Once the solder has melted around the whole joint any unmelted solder wire is moved away, followed by the soldering iron. The joint takes a few seconds to cool and harden, after which any excess wire is cut. Care must be taken to avoid dry joints, as shown in Figure 5.10.

> **Photo etching**: a printed circuit board production method where an etching solution removes copper from the board to leave the required layout.
>
> **Printed circuit board**: a plastic board with copper tracks and pads that is then populated with components to create a functioning circuit.

> **Exam tip**
>
> Make sure that you can describe how PCBs are produced and populated.

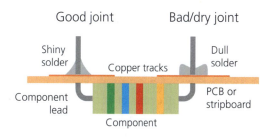

Figure 5.10 **A dry joint and a good solder joint**

## 5.6.2 Scales of production

- **One-off production** is where a single product or prototype is produced, for example a home security system made to order for a particular client.
- A breadboard is a method of building a prototype for an electronic circuit. It is a plastic board with holes for placing the components into. Underneath the plastic are rows of metal strips that make electrical connections between the holes.
- **Batch production** is where a small group of identical products are produced, for example a batch of electronic circuit boards for a particular television model.
- **Mass production** is where a product is manufactured in large quantities, usually via a production line, for example cars.
- **Continuous production** is where products are made non-stop over a period of days, weeks or even years. Components that are constantly required for use in systems, such as resistors, diodes and capacitors, are produced continuously.

> **One-off production**: production of a single, unique product.
>
> **Batch production**: production of small groups of identical products.
>
> **Mass production**: production of a very large number of products.
>
> **Continuous production**: non-stop production.

## 5.6.3 Techniques for quantity production

- **Pick and place** machines automatically select the components needed for a PCB and place them into the correct position using a robotic arm.
- SMT is where components are mounted on the surface of a PCB.
- **Quality control** involves checking a product after a process to ensure it meets the required quality standards, for example measuring a completed PCB to check it is the correct length.
- Sub-assemblies are units that are assembled separately but then fitted together with other units to produce a larger product. An example is the wheel system, chassis and power system for a robot buggy.
- It is important that tolerances are adhered to during production, for example a casing that must be cut to ± 1 mm of the specified length.
- Several techniques can be used to increase efficiency and minimise waste when cutting materials. These include tessellating and nesting, which is where shapes are fitted as close together as possible on a piece of material before cutting.

> **Pick and place technology**: machines that automatically select components for a PCB and put them into the correct position.
>
> **Quality control**: checks made to a completed product to ensure it meets quality standards.

### Marking out

- When marking out, a datum is used as a **reference point** from where all measurements are taken. A datum can also be a line or surface.
- Templates and patterns allow the same shape to be marked out repeatedly, for example, if a large number of identically shaped PCBs needs to be produced to fit inside a casing.

> **Reference point**: a point from which all measurements are taken.

### Now test yourself

1. Describe the process of correctly soldering a component to a printed circuit board. (6)
2. Explain the difference between one-off and batch production. (2)
3. State **two** methods of minimising waste when cutting materials. (2)

# 5.7 Specialist techniques, tools, equipment and processes

## 5.7.1 Tools and equipment

REVISED

- **Hand tools** are tools that are powered by the human hand. They are smaller than machinery so easier to carry around. They can be used at any time as they do not require an external power source. However, completing certain tasks with hand tools can be more time-consuming.
- Examples of hand tools used in electronic circuit construction include:
  - ○ wire cutters, which cut wires or component legs to the correct size
  - ○ wire strippers, which remove insulation from wires
  - ○ pliers, which are used for gripping or twisting wires.
- **Machine tools** make use of an electrical, mechanical or pneumatic power source. They are often static and therefore cannot be moved easily.
- Examples of machinery used in system case construction include:
  - ○ pillar drills, which create holes in a material
  - ○ scroll saws, which cut curves in wood or polymers
  - ○ milling machines, which remove material from a workpiece.
- **Digital design and manufacture** is an approach based around the use of computer systems to design and produce products. It makes use of computer-aided design (CAD) software and computer-aided manufacture (CAM) equipment.

**Figure 5.11 Wire cutters**

> **Digital design and manufacture**: an approach based around the use of computer systems to design and produce products.

## 5.7.2 Shaping

REVISED

- **Vacuum forming** can be used to shape polymers in order to create cases for systems:
  1. Firstly, a mould is placed into the vacuum-forming machine.
  2. A sheet of polymer, such as HIPS, is placed above the mould and clamped into place.
  3. The polymer sheet is heated until it becomes soft and flexible.
  4. The mould is moved upwards and air is pumped out, causing the sheet to form around the shape of the mould.
- **Laser cutters** use a high-powered laser beam to cut shapes in materials, typically polymers. The movement of the cutting head is controlled by a Computer Numerically Controlled (CNC) program that is loaded into the machine prior to cutting.
- **3D printing** is an additive process used to make prototypes of products or parts. Successive layers of material are laid down by a printer head until the completed prototype is produced. The advantage of 3D printing is that it can make complex parts relatively quickly compared to other prototyping and modelling methods.
- **Drilling** is a process in which a drill bit is used to create a circular hole in a workpiece, for example drilling small holes through the pads on a PCB for the fitting of components, or larger holes for LEDs in a casing.

> **Vacuum forming**: a process in which a heated polymer is formed around the shape of a mould through the use of a vacuum.
>
> **Laser cutting**: the use of a high-powered laser beam to cut shapes in materials.
>
> **3D printing**: an additive process where a 3D prototype is created layer by later.

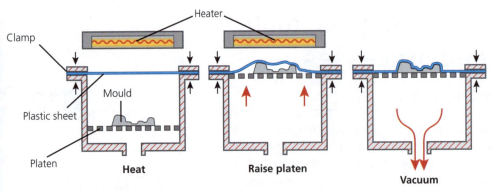

**Figure 5.12 The vacuum-forming process**

### 5.7.3 Fabricating/constructing/assembling

- PCBs can be populated using either through-hole or surface mount methods (see Circuit construction types in Section 5.5.1 on page 109).
- A **cable loom**, or harness, is a group of electrical cables that are bound together, for example through the use of plastic sleeves, cable ties or electrical tape. Looms reduce the risk of damage to cables and decrease the space needed by them.
- Wastage is the removal of material from a workpiece, for example by cutting and drilling.
- Addition is the adding of material to a workpiece, for example by soldering and 3D printing.

> **Cable loom**: a group of electrical cables that are bound together with sleeves, ties or other methods.

> **Exam tip**
>
> Make sure you can describe the use of the full range of shaping, fabricating, construction and assembly processes relevant to the production of electronic systems and their casings.

**Now test yourself**

TESTED

1. Name a hand tool used in electronic circuit construction. Describe its function. (2)
2. Give **two** advantages of 3D printing. (2)
3. Describe the purpose of a cable loom. (2)
4. Explain the difference between addition and wastage processes. (2)

## 5.8 Surface treatments and finishes

### 5.8.1 Surface finishes and treatments

Surface treatments and finishes are applied to components and systems for functional and aesthetic purposes.

### Metal plating

- Electronic connections can be **plated** with another metal in order to enhance their functionality and performance.
- For example, copper contacts can be plated with tin or nickel. Both of these materials offer good protection against corrosion. Nickel is a slightly better conductor than tin and is also very affordable.

> **Metal plating**: the coating of one metal with another to protect it and enhance its functionality.

# Insulating coatings and coverings

- Electrical wires are usually covered with a non-conductive material, such as PVC, rubber or glass. This prevents the wire inside from making contact with other conductors and creating a short circuit. It also reduces the chance of electric shocks from people touching the otherwise exposed wire.
- **Insulation** also gives a layer of protection to prevent damage to the wire itself. Insulation is often colour coded to indicate the purpose of the wire.

# Resistor colour code bands

- Most resistors have a series of coloured bands added to the outside of their casing. This is called the **resistor colour code**.
- The first two coloured bands give the first two numbers in the resistor's value. The third band gives the multiplier, for example multiply by 10, 100 or 1000. The fourth colour gives the tolerance of the resistor value. For example, gold means ± five per cent tolerance whereas silver means ± ten per cent tolerance.
- If a resistor has coloured bands of brown, black, orange and gold, it would have a value of 10,000 Ω, better written as 10 kΩ, with a ± five per cent tolerance.
- Surface mount resistors have a different way of indicating the value, which involves the use of three characters printed on the component.

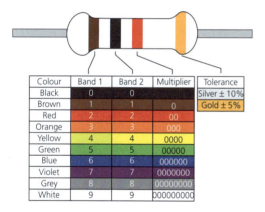

| Colour | Band 1 | Band 2 | Multiplier | Tolerance |
|--------|--------|--------|------------|-----------|
| Black | 0 | 0 | | Silver ± 10% |
| Brown | 1 | 1 | 0 | Gold ± 5% |
| Red | 2 | 2 | 00 | |
| Orange | 3 | 3 | 000 | |
| Yellow | 4 | 4 | 0000 | |
| Green | 5 | 5 | 00000 | |
| Blue | 6 | 6 | 000000 | |
| Violet | 7 | 7 | 0000000 | |
| Grey | 8 | 8 | 00000000 | |
| White | 9 | 9 | 000000000 | |

**Figure 5.14** The resistor colour code

# Finishes applied to cases

- The purpose of finishes is to improve the aesthetics of cases and to protect them from damage or corrosion.
- Metals, timbers and polymers are often painted to protect the material and to add colour.
- **Anodising** is an electrochemical process in which an oxide layer is formed on the surface of a metal, such as aluminium. This prevents further corrosion and can add decoration.
- **Screen printing** is where a design is created by pressing ink through a stencilled mesh. Although this is mainly used with textiles and papers, it can also be used with metals, timbers and polymers. It is used when particularly vibrant or vivid colouring is required.

**Insulation**: the covering of wires with a non-conductive material, such as PVC, rubber or glass.

**Resistor colour code**: a series of coloured bands that gives the value of a resistor.

**Figure 5.13** Insulated multi-strand wiring in green, red and yellow

**Typical mistake**

It is easy to misread the resistor colour code. Make sure that you understand what each of the colours mean.

**Exam tip**

Make sure that you can read and interpret the resistor colour code.

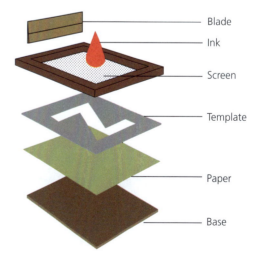

Blade
Ink
Screen
Template
Paper
Base

**Figure 5.15** Screen printing

> **Anodising**: an electrochemical process in which an oxide layer is formed on the surface of a metal.
>
> **Screen printing**: the process of creating a design by pressing ink through a stencilled mesh.

## Now test yourself

TESTED

1 Give an example of metal plating. (1)
2 Explain the purpose of wiring insulation. (3)
3 Determine the colours that would be printed on a 330 Ω resistor with a five per cent tolerance. (1)
4 Explain, using an example, why finishes are applied to materials used in system casings. (3)

## Exam practice

1 A temperature-control system must turn on a fan when it gets too hot in a room. Name a suitable sensor and output device for this system. (2)
2 State three countries from where copper is sourced. (3)
3 A 470 Ω resistor has a tolerance of ± five per cent. Calculate the maximum and minimum permissible values of the resistor. (2)
4 Write a microcontroller flowchart program that makes pin 0 high when an analogue signal falls outside of the range 100–255, but otherwise keeps it low. (4)
5 Explain the purpose of the RoHS Directive. (2)
6 Give **one** advantage and **one** disadvantage of using surface mount components instead of through-hole circuit construction. (2)
7 A 330 Ω, 470 Ω and 1 kΩ resistor are connected together in parallel. Calculate the total resistance of this arrangement. (3)
8 Describe the photo etching process. (6)
9 Explain **two** advantages of using 3D printing to produce a prototype for a circuit case. (4)
10 Give **two** reasons for anodising a metal. (2)

# 6 Textiles

## 6.1 Design contexts

- Design takes place within a context: once a need for a new product, or a problem, has been identified, the designer's job is to come up with ideas to meet that need or solve the problem.
- Emerging technologies often lead to the development of radically new products that have the potential to improve people's lives or a situation.
- For a successful outcome the designers should consider appropriate materials and their properties, applied finishes, manufacturing processes, cost and sustainability, and their moral, ethical and social implications.

## 6.2 Sources, origins, physical and working properties, and social and ecological footprint

### 6.2.1 Natural

- Natural fibres come from natural sources – animal, for example silk, and vegetable, for example linen.
- They are **sustainable** and **biodegradable**.
- Natural fibres are not as strong as synthetic fibres. They are also prone to attack from moths or insects.
- For notes on cotton and wool fibres see Section 1.11.1 Natural fibres on page 32.

Figure 6.1 Silk worms and cocoons

Table 6.1 Natural fibres

| Fibre | Source | Properties | Uses |
|-------|--------|-----------|------|
| Silk | Animal: insect | Absorbent, comfortable to wear, strong when dry, creases, good handle | Luxury clothing and lingerie, knitwear, soft furnishings |
| Linen | Vegetable: stem of flax plant | Absorbent, poor insulator, strong, cool to wear, creases easily, handles well, flammable | Lightweight summer clothing, soft furnishings, table linen |

### 6.2.2 Synthetic

- Synthetic (manufactured) fibres come from fossil fuels: oil, coal or petrochemicals.
- Synthetic fibres are **non-biodegradable** and come from **unsustainable** sources.
- Synthetic fibres are generally much stronger and more resilient than natural fibres, and more readily available.

- For notes on polyester and acrylic see Section 1.11.2 Synthetic fibres on page 32.
- Regenerated fibres are part natural (**cellulosic**) and part chemical – chemicals are used in the extraction process of the wood pulp. They have similar properties to natural fibres.
- Viscose, acetate and Tencel™ are derived from regenerated fibres.

> **Cellulosic**: fibres from a plant source.

**Table 6.2 Synthetic fibres**

| Fibre | Source | Properties | Uses |
|---|---|---|---|
| Polyamide (nylon) | Chemicals | Strong, hardwearing, melts as it burns, thermoforming, good elasticity, poor absorbency | Clothing, carpets, rugs, seat belts, ropes, tents, rucksacks |
| Elastane | Petrochemicals | Very elastic and stretchy, lightweight, strong, hardwearing | Clothing, particularly swimwear and sportswear, where stretch, comfort and fit are important |

## 6.2.3 Woven

REVISED

- Textiles can be woven using different variations with the warp and weft yarns. This will affect end use.
- For details of plain and twill weaves see Section 1.11.3 Woven textiles on page 33.

**Table 6.3 Woven fibres**

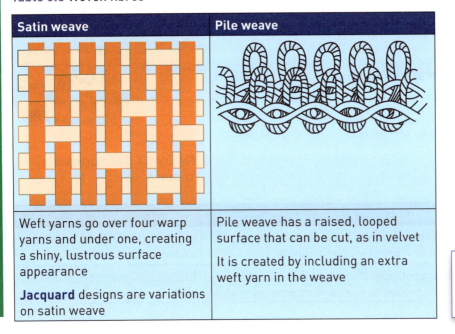

| Satin weave | Pile weave |
|---|---|
| Weft yarns go over four warp yarns and under one, creating a shiny, lustrous surface appearance | Pile weave has a raised, looped surface that can be cut, as in velvet |
| **Jacquard** designs are variations on satin weave | It is created by including an extra weft yarn in the weave |

> **Jacquard**: a geometric or floral pattern woven into the weave.

## 6.2.4 Non-woven

REVISED

For details of felted wool fabrics and bonded fibres see Section 1.11.4 Non-woven textiles on page 33.

## 6.2.5 Knitted

REVISED

For details of weft- and warp-knitted fabrics see Section 1.11.5 Knitted textiles on page 33.

## 6.2.5 Sources and origins

REVISED

**Table 6.4** Geographical sources of textile fibres

| Fibre | Origin |
|---|---|
| Acetate | Alpine forests (softwoods such as spruce, pine and hemlock; cellulose and wood pulp; **cotton linters**) |
| Cotton | USA, India, Pakistan, China |
| Linen | Russia, Canada, Ukraine, France, Belgium |
| Lyocell | European forests (hardwoods such as oak and birch) |
| Polyester, nylon, acrylic | United Arab Emirates, Saudi Arabia, Russia |
| Silk | Uzbekistan, India, China |
| Wool | UK, USA, Australia, New Zealand, China |

**Cotton linters**: fine silky fibres from the seeds of cotton plants, a by-product from cotton processing.

## 6.2.6 Physical characteristics

REVISED

- **Allergenic**: some fibres can cause minor allergic reactions when worn or used. This could be caused by their chemical make-up or through applied finishes.
- **Texture**: fabrics feel different depending on how the yarns have been made, the method of construction and applied finishes. Fabrics can be soft, fluffy, coarse or smooth.
- **Density**: fibres are spun into yarn of varying thicknesses; when woven into cloth the density of the fabric will vary too. Likewise, fabrics vary in density depending on how tightly they are woven or knitted.

## 6.2.7 Working properties

REVISED

A fabric's working properties should be considered alongside the purpose of the product to ensure that it will function as intended.

- **Tensile strength** refers to the maximum force or tension a fabric can withstand without breaking. Silk, for example, has a high tensile strength.
- **Absorbency** refers to a fibre's ability to soak up moisture or water. Cotton, for example, will absorb perspiration, keeping the wearer cool.
- Some fabrics are **breathable** as they allow air to pass through, keeping the wearer cool.
- **Electrical conductivity** is the ability to conduct electricity; conductive threads are used alongside LEDs in e-textiles.
- **Heat conductivity** is the ability to conduct heat. Heating elements are also used in e-textiles.

See Section 1.11.6 Properties on page 34 for notes on other factors to be considered, including elasticity, resilience and durability.

**Exam tip**

Designers choose materials very carefully according to their properties. This ensures the product will function as intended. Make sure you know the properties of different materials and fibres, and be able to apply your knowledge to a range of products.

**6 Textiles**

## 6.2.8 Social footprint

- The textile industry is known for being an exploitative industry with workers in developing countries often being subjected to long hours and low pay with little or no rights.
- **Fast fashion** and a **throwaway culture** have added to the exploitation of some workers.
- Work is often carried out in dangerous conditions, while those who live in close proximity to the textile industry are also affected by the contamination of land and water supplies.
- The textiles industry often causes water pollution through contaminated waste from processing and waste water from the washing, drying, dyeing and printing of fabrics.
- Many textile products are excessively packed for protection during transportation. This inevitably leads to more waste.

> **Social footprint**: a measure of the impact of an activity on society.
>
> **Fast fashion**: a recent trend copied from the catwalks that is often low quality and low in price.
>
> **Throwaway culture**: buying products for brief use then disregarding them or throwing them away with little thought of the environmental impact.

## 6.2.9 Ecological footprint

- Damage or destruction of the ecosystem is caused by clearing land to grow textile fibre crops, taking water for crop production, the building of infrastructure and pollution as a result of manufacturing and the disposal of textile waste.
- Crops such as cotton and linen are intensively farmed and require extensive areas of land and a large volume of water. Farmers use pesticides and fertilisers, which can damage the ecosystem and **biodiversity**.
- Farming livestock for wool production can lead to land being damaged by overgrazing, and from insecticides used for protection from mites and ticks. Livestock also produce the gas methane, which contributes to global warming.
- Non-renewable resources such as oil are needed for transporting raw materials and products, and for the production of synthetic fibres. This contributes to their carbon footprint.

> **Ecological footprint**: a measure of the impact that human activity has on the environment.
>
> **Biodiversity**: the variety of all the different species of organisms within an ecosystem.

### Sustainability

- An estimated 350,000 tonnes of used clothing go into landfill in the UK each year, yet most fibres and fabrics can be recycled easily. Synthetic fibres are non-biodegradable but are recyclable.
- Fibres cannot withstand endless recycling, however, and some will go to landfill at the end of their useful life.
- Crops can be farmed organically using natural fertilisers and pesticides, but this still requires a substantial amount of water.

**Figure 6.2 Overgrazing with large flocks can lead to soil damage**

### Now test yourself

1  List four synthetic fibres. (4)
2  Describe what makes regenerated fibres different to natural fibres. (2)
3  Describe **three** different ways that the textile industry is damaging the ecosystem. (3)
4  Explain why linen is suitable for summer clothing. (2)
5  Give **two** reasons why the density of a fabric can vary. (2)

> **Typical mistake**
>
> Answers relating to environmental issues must be explained. You will not gain marks by simply stating that something is good or bad for the environment. Always give a specific example in your response and a full explanation of the fact.

# 6.3 Selection of textiles

## 6.3.1 Aesthetic factors

REVISED ☐

- The appearance and form of a product is an important design consideration; failure to attract consumer interest could mean the product may not sell.
- Consumers are often attracted by image, styling, colour, **texture**, lustre and sheen on fabrics.

> **Texture**: the feel and finish of a material.

## 6.3.2 Environmental factors

REVISED ☐

- Designers aim to reduce the environmental impact of a product. Sustainable materials should be considered as well as cleaner processes during manufacture.
- Designers must also consider how the product will be disposed of when it is no longer needed.
- Fashion products can easily be **upcycled**, for example.

> **Upcycled**: unwanted products taken apart and remade into totally new products, reusing materials.

## 6.3.3 Availability factors

REVISED ☐

- Many textile materials are manufactured in stock sizes and forms. This can speed up delivery times and keep costs down.
- Specialist fabrics can be woven, dyed or decorated to order.

## 6.3.4 Cost factors

REVISED ☐

- Designers work to a budget, so the cost of raw materials will affect the selling price of a product.
- The source of the fibres also affects the cost of fabric, as do the manufacturing processes used, applied finishes and transportation of materials.

## 6.3.5 Social factors

REVISED ☐

- Fashion designers rely on predicted trends set by **fashion forecasters**. This ensures that their collections will be **on-trend** and successful.
- As different cultures and subgroups exist in modern society, different collections are developed to meet the needs of a wider target market.

> **Fashion forecasters**: people who predict future trends, often two or three years in advance.
>
> **On-trend**: up-to-date, very fashionable.

## 6.3.6 Cultural and ethical factors

REVISED ☐

- Cultural differences exist in modern society; what is acceptable to one may be offensive to another. This impacts on style of clothing and colours used depending on the culture or occasion. This differs from country to country but should always be respected.
- Many of the products bought in the UK that are used a few times then disregarded have been mass produced in factories overseas where workers are paid low wages and work in poor conditions.

- **Commercialism** drives this but organisations such as Fairtrade seek to redress the balance.
- Many products are manufactured with an intentionally short shelf life – to be used for a short period of time before being replaced or thrown away. This is referred to as **planned obsolescence**.

**Commercialism**: enterprise or business that puts profit above all else.

**Planned/built-in obsolescence**: when a product is designed to no longer function or be less fashionable after a certain period of time.

**Exam tip**

Relate a theory to a specific product to help illustrate a fact and to help explain your answer.

## Now test yourself

TESTED

1 Explain the importance of good aesthetics when designing clothes. (3)
2 Describe **one** advantage and **one** disadvantage of commercialism on people. (2)
3 Describe the importance of fashion forecasting for the fashion industry. (2)
4 Describe **two** positive reasons for upcycling old clothes. (2)

# 6.4 Forces and stresses

## 6.4.1 Forces and stresses

REVISED

Textile fabrics are subjected to different levels and types of stress when they are in use:

- **tension**: the fabric's tensile strength is its ability to resist breaking when stretched
- **compression**: the fabric's compressive strength is its ability to resist crushing
- **shear**: when fabrics are pulled against each other in opposite directions
- **flexibility**: the fabric's ability to recover after bending or twisting.

**Typical mistake**

Do not assume that a simple product such as a cushion cover is not subjected to stress or force. This could simply apply to the fibre's ability to recover after it has been leant on during use. The choice of fabric and fibre content is also important.

## 6.4.2 Reinforcement/stiffening techniques

REVISED

Designers use a number of methods to reinforce and strengthen textile fabrics where more structure is needed in the final product.

- Products such as loaded rucksacks or tents are subjected to additional forces and stresses during use. They can be reinforced with aluminium or carbon fibre frames.
- Some seams are stronger than others; for example, a double stitched seam is stronger than a plain seam.
- Quilting is a method of reinforcing and strengthening fabrics – several layers of fabric are stitched together, including an inner padded layer, for protection.
- Layering textiles, such as including a lining, will add structural support and insulation, making products more comfortable to wear.
- Piping inserted into seams adds structure to the shape of a product, helps with durability and can be decorative.
- Stay stitching is a large, straight stitch sewn in an area that could distort easily during construction. It helps keep the correct shape, for example on a curved neckline.

**Figure 6.3 Double stitched seams, top stitching, rivets and bar tacks reinforce and strengthen jeans**

**Table 6.5** Reinforcement methods

| Reinforcement method | Effect |
|---|---|
| Boning | Plastic or metal strips are sewn into reinforced seams to support the fabric, preventing creasing and buckling. Corsets that include boning accentuate and cinch the waist. Medical corsets restrict movement in people with spinal or internal injuries. |
| Rib weave | A variation on plain weave, rib weave has an extra, thicker yarn woven in at intervals to increase the strength of the fabric. This makes it abrasion and tear resistant. |
| Laminating | Layers of fabrics can be glued or bonded together to improve functionality, aesthetics and strength. Neoprene, a synthetic rubber used in wetsuits, is often laminated to a knitted polyester for functional reasons. |
| Interfacing | A non-woven, bonded fabric that can be sewn or ironed on to the outer fabric or lining. It provides support, stiffens and reinforces fabrics. |
| Composite materials | Reinforced composite materials are lightweight, strong and impact resistant. Kevlar® is a composite material that is five times stronger than steel and resistant to knife attack. |

## Now test yourself

TESTED ☐

1 Explain the purpose of boning in a medical corset. (3)
2 Give **one** reason for using stay stitching on a shaped neckline. (1)
3 List **three** places that interfacing would be used on a woven school shirt. (3)
4 Give **one** detailed reason for laminating fabric together. (2)
5 Explain why quilted panels are often used on heavy-duty workwear. (2)

**Exam tip**

Almost all textile products have some sections that are reinforced or strengthened to improve their structural integrity and functionality. The more products you analyse during your studies the better your understanding will be. Use this information when answering exam questions.

# 6.5 Stock forms, types and sizes

## 6.5.1 Stock forms/types

REVISED ☐

Textile fabrics are readily available in a variety of stock forms:

- Fabric is sold off the roll (or **bolt**) by the metre.
- Designers and pattern technologists work from standard block templates that are manipulated and cut to create new templates for original designs.
- Denier refers to the fineness of a fabric and is indicated by a number; for example, 100 denier tights would be quite thick whereas 10 denier are fine.
- Fabric weight varies according to the construction method, fibre type and applied finishes.
- Most fabrics are single weight but those that appear the same on both sides are heavier double weights. These can be woven or knitted and are usually more expensive.
- Laminated fabrics are also available in stock forms in various weights, for example fabrics used in outdoor performance clothing such as Gore Tex.

**Bolt**: a short, flat roll of folded fabric.

**Figure 6.4** Textile fabrics are available in a range of fibre content, colour, texture and weight

**Exam tip**

Know the common names of textile fabrics, such as corduroy, chiffon and cotton poplin, and not just their fibre source. This demonstrates a higher level of understanding. Cotton, for example, is the name of the fibre – there are many different variations when made into fabric.

### 6.5.2 Sizes

- Standard widths for textile fabrics include 90 cm (interfacings or linings), 137 cm and 154 cm.
- Cotton sheeting at 240 cm wide is also available from some specialist fabric retailers.
- Fat quarters are pre-cut fabric pieces measuring 45 cm × 55 cm, ideal for small projects.
- Yarns are classed according to their thickness and the number of threads that are spun together to make the yarn: 4 ply, for example, has four threads, while 2 ply is much thinner.
- Different widths of fabrics should be considered to find the best use of fabric. Pattern templates must be laid out correctly to minimise the amount of waste fabric.

### Calculations

Surface area of a rectangle = width × length

Surface area of a circle = $\pi r^2$

Circumference of a circle = $2\pi r$

---

**Now test yourself**    TESTED

1. Each edge of a cube-shaped floor cushion measures 40 cm. Calculate the surface area of the cube. (3)
2. Fabric denier is indicated by a number. State what a high number indicates. (1)
3. Explain what is meant by a double weight fabric. (2)
4. List the three stock forms for fabric widths. (3)
5. Explain what a 3 ply yarn is. (2)

---

**Typical mistake**

Where calculations are needed, such as working out costs, show all of your workings even if you have used a calculator. You will lose marks if you do not show the method used.

---

# 6.6 Manufacturing to different scales of production

## 6.6.1 Processes that can be used to cut and shape materials

- Fabric shears have long blades to cut fabric easily.
- Laser cutters controlled by computer-aided design (CAD) software can cut or engrave complicated patterns into fabric. In industry, lasers are used to cut through multiple layers of fabric.
- Band saws are used in industry to cut through large numbers of layers of fabric accurately and quickly.
- Automated die cutters (stamps) are used to cut consistent shapes through several layers of fabric.

**Figure 6.5** Fabric shears

- Soldering irons are tools that contain a heated element. They can be used to melt and seal the edges of fabric that is prone to fraying.
- Some synthetic fibres are extruded (polymers in liquid state are forced through tiny holes in a spinneret) into long, continuous filaments before being spun into yarn.

## 6.6.2 Scales of production

### One-off, bespoke or job production

- **One-off**, custom-made or **bespoke** products are made by an individual or small team of highly skilled, versatile workers.
- One-off products are often commissioned as unique pieces by clients who may also have direct input into the design process. They often use high-quality fabrics and components, and are expensive, for example haute couture and designer gowns.

> **Exam tip**
>
> The prototype product you make for your non-examined assessment could be considered a one-off design. It should be immediately useable, fully functional and made for an individual. Keep that in mind when thinking about scales of production.

### Batch production

- **Batch production** is used to produce a specific number of identical products in a set timescale, such as seasonal fashion products. They are usually of mid to low quality.
- The products are made by large teams who specialise in one element of the construction process.
- Batch production is flexible and can change to meet market demand. Repeat orders can be facilitated quite easily.

### Mass production

- **Mass production** is usually used for products that are in high demand over a long timescale.
- Workers are skilled or specialised in one element of the process.
- Sub-assembly lines usually complete part of the main product.
- Typical mass-produced products include socks and plain T-shirts, where styles rarely change.

### Continuous production

- High-volume production that continues 24/7 over extended periods of time.
- High levels of automation are often used to maintain productivity.
- Synthetic fibres are manufactured using continuous production.

> **One-off**: a single, unique product.
>
> **Bespoke**: specially made for a particular person.

**Figure 6.6 A bespoke wedding dress being made for an individual client**

> **Batch production**: a limited number of identical products made in a set timescale.
>
> **Mass production**: very large numbers made continuously over long periods of time.

> **Typical mistake**
>
> Be clear on the differences between batch and mass production. Descriptions that are vague or too similar will not gain credit.

## 6.6.3 Techniques for quantity production

It is important that consistent approaches are used to ensure the quality of the final products.

- Templates and **patterns** have important markings or instructions that indicate how they should be laid on the fabric, what the pieces are, how many are needed and any details such as button placements. Some marking will be transferred on to the fabric.
- **CAD** software is used to develop digital lay plans, allowing manufacturers to tessellate the pattern pieces to maximise fabric usage. These programs are linked to automated cutting machines.
- **Computer-aided manufacture (CAM)** is increasingly used in mass production where quality, consistency and speed are important, for example multi-head embroidery machines for embroidered details.
- **Quality control** is a system of checks that ensure all products meet the required standard and are equal in terms of quality, size and shape.
- Some textile products are more complex, so a permitted **tolerance** of a seam of about ± 1 cm is acceptable. Where there are a number of processes involved, 100 per cent accuracy is almost impossible.

> **Patterns**: shapes representing parts of a product that are placed on fabric and cut around before being joined together to make the product.
>
> **Tolerance**: the allowable amount of variation of a stated measurement.

### Now test yourself

TESTED

1 Give a reason why seasonal products might be batch produced. (1)
2 Describe an advantage of having a bespoke item of clothing made. (2)
3 Describe a disadvantage of working on a mass assembly production line. (2)
4 List **four** factors that influence the scale of manufacture. (4)
5 Describe the importance of quality control during the manufacture of clothing. (3)

# 6.7 Specialist techniques, tools, equipment and processes

## 6.7.1 Tools and equipment

### Hand tools

- Metre rule for marking out and cutting fabric, and measuring hemlines.
- Craft knife for cutting small templates and stencils.
- Quick unpicker to unpick faulty seams.
- Tracing wheel and carbon transfer.
- Hot notch markers.
- Pinking shears with a zig-zag cutting edge that prevents fraying.

- Pins to hold fabric together temporarily.
- Cutting mat with guides for accurate cutting; used with a rotary cutter or craft knife.
- Tape measure for taking body measurements.
- Tailors chalk.
- Fabric shears for cutting out fabric.
- Embroidery scissors for cutting intricate work.

## Sewing machines

- Domestic sewing machines have a wide range of facilities and stitches to complete different processes including optional feet for different processes such as attaching zips.
- Sewing machine needles vary and should be selected according to the fabrics being used.
- Computerised sewing machines can embroider original designs.
- An overlock machine is used in industry to cut a straight edge on the material and over sew the edge to neaten the seam in one process.

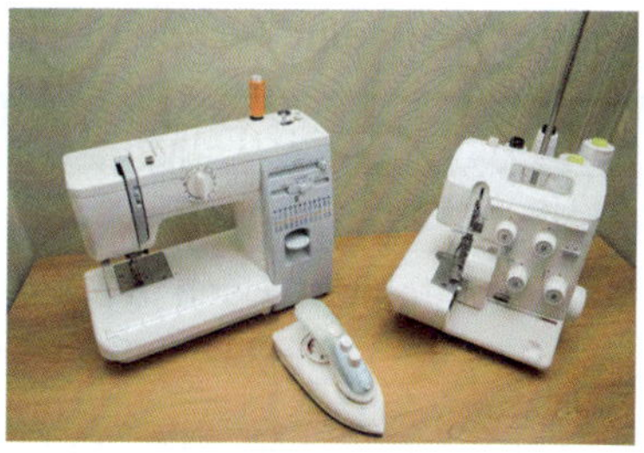

**Figure 6.7** Domestic sewing machine, overlocker and iron

## Laser cutting

- Laser cutting is controlled by a CAD program. The design is drawn as a 2D image that the laser follows to accurately cut or etch the design. The laser strength and speed depend on the material to be cut.
- Disadvantages include:
  - not all fabrics can be cut as some melt and burn
  - the laser leaves an unsightly burnt brown edge on some fabrics.

## Digital design and manufacture

- Designers use CAD to develop surface designs for printing that can be transferred to relevant digital printing systems.
- Digital designs can be revisited, manipulated and changed, including the development of patterns and colourways.
- CAD is a more cost-effective way of developing designs. By using 3D imagery or virtual prototyping, the need to make numerous prototypes is reduced. It is considered a more sustainable way of designing.
- Computers can control sections of manufacturing in the textile industry. Some machinery is semi-automated, requiring some human input, while other machines are fully automated.
- CAM systems are expensive but speed up manufacturing, improve productivity and consistency, and reduce human error.
- CAM applications include multi-head embroidery machines, digital printing on fabrics, laser cutters, 3D printers and automated fabric spreaders.

> **Typical mistake**
>
> CAD and CAM may offer speed to complete processes, but they offer much more than that. Marks are often missed when answers are not explained.

## 6.7.2 Shaping

REVISED

**Table 6.6** Shaping techniques

| Technique | Description |
| --- | --- |
| Pleats | The fabric is folded back on itself and sewn in place, narrowing the width of the material but adding shape. Pleats can be used as decorative frills. |
| Tucks | Similar to pleats to draw in fullness. |
| Gathers | The fabric edge is gently drawn in to reduce and narrow the original width of the fabric, giving fullness and shape to a product. Gathers can also be decorative. |
| Darts | Made by creating folds in the fabric that taper to a point to improve shape and fit. Most commonly used around the bust area. |
| Shirring | Several parallel rows of very narrow corded elastic are stretched and stitched in place to gently draw in fullness. |
| Ease | A stitch is slightly pulled up smoothly, drawing in some fullness and giving a rounded shape. |

**Table 6.6 Shaping techniques (continued)**

| Technique | Description |
|---|---|
| Godet | A triangular panel is inserted into a seam to give a flared effect. |
| Under stitching | Used to prevent a facing or lining from showing above the edge of the outer fabric. The facing is stitched down to the seam allowance, close to the edge of the fabric. |

- Some fabrics can be moulded into shape by applying heat, steam or adhesive. Thermoforming polymers can be heat set and moulded or pleated into a set shape.
- For adding structure through boning and interfacing see Section 6.4.2 Reinforcement/stiffening techniques on page 123.

### 6.7.3 Fabricating/constructing/ assembling

REVISED ☐

Some designers start designing and constructing a garment by draping their chosen fabric over a dress form. This allows them to assess the fabric's suitability, as well as to judge proportion.

### Seams

- The type of seam used will depend on the product and fabric. It is important to use the most suitable method.
- The standard **seam allowance** is 1.5 cm. This must be applied consistently if the product is to fit together as intended.

**Table 6.7 Different seams**

| Seam | Description |
|---|---|
| Plain seam | Most common seam type<br>Seams can be stitched together or pressed flat open<br>Double stitched seams are a variation with two rows of stitching |
| French seam | French seams are enclosed and hide all **raw edges**<br>Used on more expensive clothing and sheer materials where seams need to be hidden |
| Flat felled seam | Strong seams with two rows of stitching adding to strength<br>**Top stitching** is often in a different colour<br>Lapped seams look similar to flat fell seams |

**Figure 6.8 Pleats are stitched close to the waistband**

**Figure 6.9 A designer drapes the fabric over the dress form as part of the design process**

**Seam allowance**: distance between the edge of the fabric and stitching line.

**Raw edges**: unfinished edges; the cut edge.

**Top stitching**: a visible line of stitching with a thicker thread, often in a contrast colour.

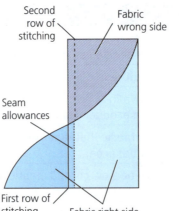

**Figure 6.10 Plain seam**

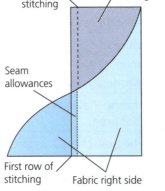

**Figure 6.11 French seam**

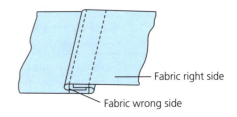

**Figure 6.12 Flat felled seam**

## Finishing raw edges

Edges are finished to prevent fraying. Methods include:

- **Overlocking**, which simultaneously joins the seam while trimming off excess fabric along the seam; the stitch overlaps the edge. An overlocked edge can have two, three or four threads depending on the finish required.
- **Zig-zag stitch** sewn along the raw edge of a seam.
- Edges can be bound with **bias binding**.
- A **rolled hem** is very narrow, fine gauge overlocked stitch.
- Hems can be finished by folding up the edge of the fabric and either machine stitching the edge in place or using a **blind hem stitch** or **invisible stitching**.
- On a folded hem, a double-sided fusible strip can be sandwiched inside the folded edge holding the hem in place.

In garment construction all seams should be carefully pressed as an ongoing process to allow a high-quality finish.

## Fusing

Different seams are needed on waterproof products as moisture can seep through the holes created by sewing needles. Tape is bonded to the seams to seal them, making the seam watertight.

## Other methods of fabrication, construction and assembling

- Component linkage refers to the coupling together of component parts to join pieces, for example a toggle and corded loop used on fabrics as a fastening.
- Set shaped fabric pieces can be cut in fabric using a stamp – the shaped cutter is pressed into the fabric to cut the desired shape.
- Some woven and non-woven fabrics can be moulded into 3D shapes using steam, heat or adhesive, for example felt hats can be moulded to a person's head.

## Wastage

Manufacturers seek to reduce wastage with more effective lay planning of templates. Additional waste is created through the disposal of clothing that could easily be recycled.

## Addition

Addition in textiles refers to an additional piece of fabric being added, for example in quilting where layers are added together.

**Blind hem stitch**: a stitch that results in hems that are invisible from the outside.

**Invisible stitching**: concealed stitching that is not visible on either side of the fabric.

**Figure 6.13** Seam finished with a zig-zag

**Figure 6.14** Edge finished with bias binding

### Exam tip

Carefully consider what the function and purpose of a product is. This will help you decide what essential properties the fabric should have and help you to identify a suitable fabric and construction method.

## Now test yourself

TESTED ☐

1. Explain why a French seam would be used on sheer fabric. (2)
2. Describe **two** reasons for using an overlocker in garment construction. (2)
3. List **three** different methods of finishing a raw edge in a seam. (3)
4. Explain why stay stitching would be used during garment construction. (2)
5. Give **one** reason for using a blind hem stitch. (1)

# 6.8 Surface treatments and finishes

## 6.8.1 Surface finishes and treatments

Finishes are applied to fabrics for different reasons, for example to improve aesthetics, comfort or function. They are applied mechanically, chemically or biologically. Some surface finishes and treatments are known as embellishments or **decorative techniques** and are applied for aesthetic reasons.

> **Decorative techniques**: used to improve the aesthetics of a product by adding colour, texture and pattern, for example dyeing, printing and embroidery.

### Painting

- Fabric paints, fabric felt pens and pastel crayons are applied directly on to textiles. Once dry, the paint should be fixed with a hot iron.
- Specialist silk paints give a delicate watery effect and can be used with Gutta outliner, which acts as a barrier to separate sections of the design.

### Dyeing

- Dyeing is the most common method of colouring fibres and fabrics.
- Natural dyes work well on natural fibres, while synthetic dyes will give deeper or brighter colours. Synthetic fibres need chemicals to take on the dyes.
- Methods of dyeing textile fabrics include:
  - **Tie and dye**: the fabric is tied or knotted to produce unique and varied designs. This is a **resist method** of colouring.
  - **Batik**: also a resist method where hot, melted wax is applied to fabric in the desired pattern. Once cooled the fabric is submerged in a dye bath or the dye can be painted on directly.

**Figure 6.15 A hand-painted silk scarf – the white outline that separates the flowers is created with Gutta outliner**

> **Resist method**: a means of preventing dye or paint from penetrating an area on the fabric to create patterns.

### Printing

- The **silk screen printing** (flatbed printing) process is as follows:
  1 Fine mesh fabric is stretched over a wooden frame.
  2 Part of the screen is masked out.
  3 Printing ink is placed on the underside of the frame.
  4 A squeegee is used to drag the ink across the screen, forcing the ink through the fabric to leave the design on the material.
- **Rotary/roller printing**: similar to silk screen printing but uses metal cylinders instead of screens.
- **Stencilling**: stencils are usually made from a thin sheet of card or plastic with a pattern cut out. Paint or dye is applied through the holes in the stencil to leave a design on the fabric.
- **Block printing**: paint or dye is applied to a relief block that is pressed on to fabric to create a pattern. Repeating the process creates a repeat pattern.
- **Digital fabric printing**: large inkjet printers and specialist dyes transfer a digital image to the fabric. It can be used to create intricate and detailed images, and to test sample pieces before production.

## Embroidery

- **Machine embroidery**: in-built decorative stitches that can be used to enhance any design.
- **Free machine embroidery**: the fabric is secured in a frame and the machinist moves the fabric around freely to create a design.
- Computerised sewing machines either have preinstalled designs or are linked to CAD packages to create and stitch original designs.
- **Appliqué**: a traditional way of applying a design by stitching separate pieces of fabrics on to a base fabric.

## Quilting, patchwork, couching

- **Quilting**: created by layering fabrics: a top decorative layer; a thicker middle layer, for example wadding; and a backing fabric. These are stitched together in a decorative design.
- **Patchwork**: consists of several small pieces of fabric sewn together.
- **Couching**: a thick cord is machine-stitched or hand sewn on to a base fabric to give a raised effect.

## Applied finishes

### Chemical finishes

- **Mercerising**: caustic soda is used to make cellulosic fibres in the fabric swell, leaving a more lustrous and stronger fabric.
- **Crease resistance/easy care**: a resin coating is applied to the fabric, stiffening the fibres and making products easier to care for. Absorbency is reduced, however.
- **Flame resistance/fire proofing**: chemicals such as Proban® are applied to the surface of fabrics as a liquid coating, which reduces the fabric's ability to ignite and burn.
- **Bleaching**: removes any natural colour; it is used to prepare fabric for dyeing and printing.
- **Stain resistance**: Teflon™ and Scotchguard™ are fabric protectors mostly used on clothing and home furnishings.
- **Anti-static**: a chemical-based product is applied to prevent the build-up of static electricity.
- **Water repellence/water proofing**: silicon is applied as a semi-permanent finish that repels water. Applying a fluorochemical resin makes fabric water repellent and wind resistant. Teflon™ and Scotchguard™ are water repellent finishes. Coating with PVC, PVA or wax also repels moisture.
- **Shrink resistance**: a chlorine-based chemical smooths out the scales on wool fibres to prevent them locking together during washing, preventing shrinkage.
- **Carbonising**: removes any plant fragments from newly woven fabric.

### Mechanical and physical finishes

- A smooth surface is achieved by **singeing** the fabric by passing it through a naked flame to remove excess fibres. It is then **desized** by treating it with an enzyme or chemical solution to remove excess starch, improving its draping qualities.

**Figure 6.16** A modern interpretation of an appliqué design

- **Heat setting** the fabric allows warp and weft yarns to be realigned following construction.
- A raised surface is achieved either by brushing the fabric with wire brushes or **emerising** it – brushing the surface of the fabric with sandpaper.
- **Calendering** involves passing the fabric through heated rollers to give it a smooth, more lustrous surface, enhancing its aesthetic qualities. Engraved rollers give an embossed effect.
- **Milling** and **fulling** increase the thickness and compactness of yarns, making them easier to work with.

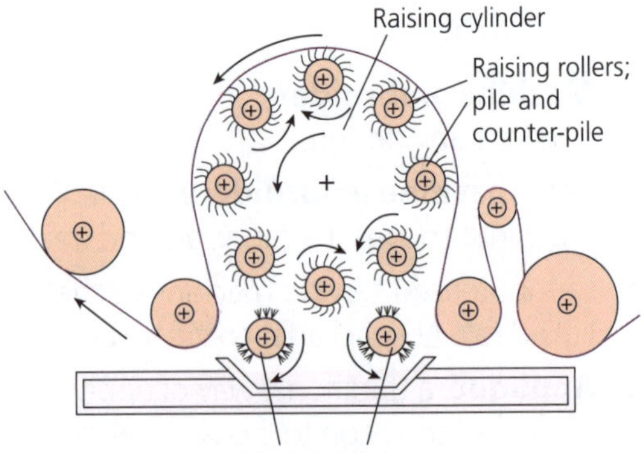

**Figure 6.17 Wire brushes raise the fibres to produce a soft, fluffy surface**

### Biological finishes
Enzymes are used to modify fabrics:

- **Biostoning** uses enzymes that act on the dye to create a washed-out look.
- **Biopolishing** uses enzymes to remove excess fibre, giving a smoother appearance.

## Smart
Smart fibres and fabric react to external stimuli:

- **Solvation chromism**: these materials react and change colour in response to varying moisture levels, for example disposable nappies.
- **Electrochromic**: these materials respond and change colour in response to an electric charge. Electrochromic materials can be woven into fabrics.

For photochromic, thermochromic, antibacterial and micro-encapsulation, see Section 1.4 Modern and smart materials, composite materials and technical textiles on page 16.

**Exam tip**

To demonstrate high levels of knowledge and understanding, know how finishes are applied to fabrics and be able to explain the purpose of the finish, particularly in relation to the end product.

**Typical mistake**

Show understanding of the stages needed to apply surface treatments and finishes to fabrics. Stages in a process need detail to fully explain them. Marks may not be awarded for a sequence that lacks detail, is incomplete or in an incorrect order.

### Now test yourself
TESTED

1 Describe the process of appliqué. (2)
2 Describe how the process of tie and dye works. (3)
3 Name the finish for reducing electrostatic charge in textile fabrics. (1)
4 Give **two** reasons for applying a Teflon™ coating to textile fabrics. (2)
5 Name the process and explain how a washed-out look is achieved on textile fabrics. (3)

## Exam practice

1 Weaving and knitting are the two main methods of fabric construction.
   a Describe **two** reasons for weaving fabric in different ways. (2)
   b Explain why knitted fabrics are often used in casual leisurewear. (2)
   c A rib weave is a variation on a plain weave. Explain how this weave affects the fabric's properties. (3)
2 Fibres are the raw materials of textiles.
   a Underline the **two** synthetic fibres in the list below. (2)
     silk          polyester          acrylic          linen
   b Explain why silk is a suitable fibre for luxury clothing. (2)
3 The textile industry has a major impact on the environment.
   a Explain why synthetic fibres have a negative impact on the environment. (3)
   b Describe the impact that cotton fibre crops have on the environment. (4)
   c Explain the effect that the transportation of textile goods has on our carbon footprint. (3)
   d Explain the impact a throwaway society has on:
     i   workers in developing countries (2)
     ii  the environment. (2)
4 Finishes are applied to textile fabrics for different reasons.
   a Explain the purpose of a Scotchguard™ finish. (2)
   b Explain how brushing enhances the functionality of cotton fabrics. (3)
5 Components are used in textile products for a range of different reasons.
   a Give **two** reasons for using piping as an edge finish on a textile product. (2)
   b A textile student has decided to make a circular floor cushion that will have a piped edge inserted into the seams of the two circular end pieces, as shown on the diagram below.
   The diameter of each circular end piece is 70 cm.
   The height of the floor cushion is 30 cm.
     i   Calculate how much piping cord will be needed to complete the cushion. (4)
     ii  The sides of the floor cushion will be cut from one rectangular piece of fabric. Calculate the size of the template that will be needed to form the side of the floor cushion. Include a seam allowance of 1.5 cm in your calculation. (2)

Piped edge

6 Fashionable clothing is made using different scales of production.
   a Explain why batch production would be a suitable scale for the manufacture of children's woollen coats. (3)
   b Describe the benefits to the client of having a bespoke product made. (4)
7 Analyse the use of computer-aided design (CAD) in lay planning in garment product. (5)
8 Decorative techniques are used throughout fashion and textiles.
   a Place a tick in the correct boxes in the table below to show whether the statements are true or false. (4)

| Statement | True | False |
| --- | --- | --- |
| Natural dyes work best on synthetic fibres | | |
| Roller printing is another name for flatbed printing | | |
| Free machine embroidery allows the fabric to move freely while being stitched | | |
| Batik involves using hot, melted wax to outline a shape on fabric | | |

   b Evaluate the use of CAD in the design and development of digital printing on textile fabrics. (6)
9 Textile products are manufactured using different methods to join fabrics.
   Explain why a flat fell seam would be used in the construction of a hot air balloon. (4)
10 Different construction techniques are used during the manufacture of clothing.
   a Explain what a dart is and why it would be used on a dress. (3)
   b Explain the purpose of gathers being used at the waist of a skirt. (3)

# 7 Timbers

## 7.1 Design contexts

Timber is a strong, lightweight, decorative and, most importantly, renewable material. Timber is used in many applications, such as construction, furniture making, kitchen utensils, artworks and sports equipment. Many different types of timber and manufactured boards are available to the designer and manufacturer. It is important to know their properties so that you can make an informed judgement as to which to use.

## 7.2 Sources, origins, physical and working properties, and social and ecological footprint

### 7.2.1 Natural timbers – hardwoods (including 7.2.4 Sources and origins, 7.2.5 Physical characteristics and 7.2.6 Working properties)

REVISED ☐

Hardwoods come from slow-growing, deciduous trees. Deciduous trees can be identified easily as they have broad leaves that are lost in winter, and they carry their seeds within fruit.

**Table 7.1** Hardwoods

| Hardwood | Geographical location | Properties | Common uses |
|----------|----------------------|------------|-------------|
| Oak | European forests | A hard, tough, strong, durable timber with an attractive light, open grain containing few knots; hard to work; can be finished to a high standard; has natural weather protection | Timber-framed buildings, high-quality furniture, flooring, artworks, wine and whisky barrels |
| Mahogany | Amazonian rainforests | A strong and durable timber with a deep reddish colour containing very few knots; available in wide planks; fairly easy to work but can have interlocking grain; has some natural protection to weathering but can accept a high-quality finish | Good-quality furniture, panelling, veneers, window and door frames |
| Beech | European forests | A hard, strong, close-grained timber; has a light brown colour with distinctive flecks of brown and contains very few knots; prone to warping and splitting; can be difficult to work; needs to be protected from weathering | Furniture, children's toys, workshop tool handles and bench tops |

**Table 7.1** Hardwoods (continued)

| Hardwood | Geographical location | Properties | Common uses |
|---|---|---|---|
| Balsa | Central and South American forests | A very lightweight, soft and easily worked timber that contains very few knots; pale in colour but weak and not very durable; very poor natural resistance to weathering | Model making, floats and rafts |
| Jelutong | Asian forests | A close-grained timber with a pale colour that contains very few knots; medium hardness and toughness; easily worked; very poor natural resistance to weathering | Pattern making |
| Birch | European forests | Even, straight-grained timber that has a medium strength, is easy to work but contains some knots; attractive 'clean', light-coloured grain; poor resistance to weathering | Veneers, plywood, internal doors and furniture |
| Ash | European forests | Strong, tough, durable and flexible with a few knots; has an open, attractive, light-coloured grain; finishes well; some natural resistance to weathering | Handles for tools, sporting equipment, furniture, ladders |

# 7.2.2 Natural timbers – softwoods (including 7.2.4 Sources and origins, 7.2.5 Physical characteristics and 7.2.6 Working properties)

REVISED

Softwoods come from relatively fast-growing coniferous trees. Coniferous trees have needles that they keep all year round and they carry their seeds in cones. Larch trees are one of a few softwood trees that do not keep their needles all year round.

**Table 7.2** Softwoods

| Softwood | Geographical location | Properties | Common uses |
|---|---|---|---|
| Pine | Alpine forests | A straight-grained, light yellow-coloured timber with a distinctive grain; strong and lightweight; soft and easy to work; can be quite knotty; needs to be protected from weathering | Interior joinery and furniture, window frames, roof trusses and flooring |
| Cedar | Alpine forests | Very resistant to weathering and decay; has a light reddish-brown colour with a close, straight grain containing few knots; easily worked | Fencing, fence posts and exterior cladding |
| Larch | Alpine forests | Tough, durable, good weather resistance; has a light-coloured grain containing some knots | Exterior cladding, fencing, fence posts, sheds |

**Exam tip**

Make sure you can name several different types of timber, be able to classify them, and give their properties and uses.

**Typical mistake**

Candidates often confuse hardwoods and softwoods.

# 7.2.3 Manufactured timber (including 7.2.4 Sources and origins, 7.2.5 Physical characteristics and 7.2.6 Working properties)

**Manufactured boards** are processed from many different types of natural and recycled timber. They have many advantages over natural timber – they are less likely to twist or warp, and are available in much larger sizes.

> **Manufactured boards**: human-made boards that are available in large flat sheets.

Table 7.3 Manufactured timber

| Manufactured board | Description | Properties | Uses |
|---|---|---|---|
| Plywood | Made from wood veneers glued together with an alternating grain | Very strong, flat, smooth surface; resistant to twisting and warping; outside veneer can be made from a wide range of hardwoods; most plywoods need protection from weathering; relatively easy to work but prone to splintering | Construction and furniture Boat building |
| Medium-density fibreboard (MDF) | Made from compressed fine wood fibres bonded together with resin | Relatively inexpensive and has a flat, smooth surface; has very little protection from weathering; easy to work but is abrasive on tools; suitable for machining; its dust is a health hazard | Inexpensive flat-pack furniture, floor coverings, wall panelling |
| Chipboard | Made from wood chips bonded together with resin | Inexpensive construction material with limited strength; has a flat but textured surface; has very little protection from weathering; easy to work but is abrasive on tools; suitable for machining | Kitchen work tops, floor coverings, flat-pack furniture |

# 7.2.7 Social footprint

## Trend forecasting

- Trend forecasting is concerned with predicting the use of timber products over the next few years. This can affect the supply and demand of natural resources.
- New, affordable housing is utilising the concept of offsite timber construction. This will lead to an increase in the need for straight-grained softwoods.
- The use of tropical hardwoods from the Amazon rainforest is harming our planet so there is likely to be less demand for this type of timber.

## Impact of logging on communities

- Logging involves the **felling** (cutting down) of trees and their transportation to sawmills for conversion into planks.
- This can provide employment for communities and involves the building of roads and infrastructure.
- In poorer countries logging can lead to the loss of towns and villages, and the destruction of fertile ground, if trees are not replanted.
- The **biodiverse** nature of forests can be lost, and wildlife and plant life may suffer.

> **Felling**: the process of cutting down trees.
>
> **Biodiversity**: the variety of all the different species of organisms within an ecosystem.

## Recycling and disposal

- Timber products can be **upcycled**, giving them a life as another product once their first life has been fulfilled. Railway sleepers can be reused to build flower beds, for example.
- Some timbers can be recycled into manufactured boards such as chipboard and MDF.
- Waste timber can be used as a fuel at a biomass power station.
- If natural timber goes to landfill, it is fully biodegradable and will rot without harming the soil.
- Manufactured boards contain resins and are far less easy to recycle or dispose of.

> **Upcycled**: unwanted products taken apart and remade into totally new products, reusing materials.

## 7.2.8 Ecological footprint

REVISED ☐

### Sustainability

- Natural timber is renewable and, if managed correctly, we need never run out.
- The Forest Stewardship Council® (FSC®) ensures that forests are managed, meaning that trees are replanted.
- Timber products are relatively easy to repair; therefore, their life can be extended if they break.

### Deforestation

- **Deforestation** occurs when trees are cut down and not replanted.
- Fertile soil is washed away, silting up rivers and flooding land.
- Trees convert carbon dioxide into oxygen – if they are not replaced, the carbon dioxide level in the atmosphere increases, adding to global warming.

> **Deforestation**: the large-scale felling (cutting down) of trees that are not replanted.

### Habitat destruction and loss

The forest is home to many species of plants and animals. If the forest is lost, they will have nowhere to live and may even become extinct.

### Processing

- The **conversion** (sawing) of logs into planks involves mechanised saws that burn fossil fuels.
- Air **seasoning** is a natural method of drying out timber that does not harm the planet.
- Kiln seasoning uses energy, which creates carbon dioxide that is released into the atmosphere.

> **Conversion**: cutting trees into planks.
>
> **Seasoning**: drying out newly felled timber.

### Transportation

- The most ecological way to transport logs to the sawmill is by floating them down rivers.
- Lorries burn fossil fuels, creating toxic gases that pollute the atmosphere.

## Wastage

- Waste from the processing of trees into planks can be dealt with in an ecological way.
- Branches and poor-quality timber can be processed into manufactured boards.
- Smaller branches and bark can be used as fuel at a biomass power station.
- As timber is biodegradable, it will rot down without harming the soil if sent to landfill.

## Pollution

- If wood is burnt to clear land, it releases carbon dioxide into the atmosphere.
- The transportation and processing of timber uses fossil fuels that release toxic gases into the atmosphere.

# 7.3 Selection of timbers

## 7.3.1 Aesthetic factors

REVISED ☐

### Form

- Timber structures are usually made up of straight lines, as timber is initially cut into straight planks.
- Timber can be cut and shaped to produce 2D curves.
- Timber can be bent into 3D shapes by the processes of laminating and steam bending.

### Colour and texture

- The natural colour of timber can vary from very pale (sycamore) to jet black (ebony).
- Timber such as ash and pine has a very pronounced grain.
- Timber can be painted and stained to change its colour.
- The surface of timber can be left rough or sanded and polished to a very smooth finish.

**Figure 7.1 A wooden bowl shows the effect of colour and grain on a piece of timber**

## 7.3.2 Environmental factors

REVISED ☐

- **Sustainability**: timber is a renewable resource and, if we look after it, we can use timber indefinitely.
- **Genetic engineering**: scientists can alter the DNA of trees to improve their properties. In China, poplar trees are being modified to make them resistant to attack by insects.

> **Genetic engineering**: when the DNA of timber is altered to improve its properties.

- **Seasoning**: the moisture content of newly felled timber needs to be reduced to match its surroundings. Seasoning timber turns it into a workable material and helps prevent defects such as twisting, warping and attack by insects and fungus.
- **Upcycling**: when a timber product reaches the end of its life, instead of going to landfill it can be upcycled into another product. For example, wooden pallets can be made into garden furniture.

## 7.3.3 Availability factors

REVISED

- **Stock materials**: timber is available in many stock sizes. Designers and manufacturers can significantly reduce the overall cost of the project by using stock sizes. For more information see Section 7.5 Stock forms, types and sizes on page 142.
- **Specialist materials**: marine plywood is a **water-resistant** manufactured board used for building wooden boats. Exotic hardwood veneers such as burr walnut can be used to enhance the appearance of luxury car dashboards.

> **Water resistance**: ability to resist penetration by water or moisture.

- **Hurricanes, storms and diseases**: high winds can bring down trees, and diseases such as Dutch elm disease can wipe out an entire species.

## 7.3.4 Cost factors

REVISED

### Quality of material

- Timber is sorted, graded and priced by quality. High-quality timbers will have fewer defects, such as knots and splits.
- Plywood is graded from A to D, with A being the best quality – it is free from defects and has an even colouring.

### Manufacturing processes

Each process that is carried out on timber has an inevitable cost. Rough-sawn softwoods are generally inexpensive as they have had the least amount of processing. Planed timber is more expensive.

### Treatments

Each treatment that is applied to timber increases its cost. **Tanalised** timber has been pressure treated with a preservative to make it suitable for use as patio decking. Fire-retardant paints and coatings can be applied to timber to help slow down the spread of fire in timber-framed buildings.

> **Tanilising**: pressure treatment of timber with preservatives to improve its resistance to weathering.

## 7.3.5 Social factors

REVISED

- **Different social groups**: furniture produced for the mass market is made from affordable manufactured timber such as MDF. Exotic timbers such as walnut and mahogany are expensive so are generally used to produce bespoke furniture for wealthy clients.
- **Trends, fashions and popularity** affect the choice of timber demanded by customers. In the 1950s, affordable hardwood furniture became popular, while in the 1980s black ash-veneered chipboard was in vogue.

## 7.3.6 Cultural and ethical factors

- **Avoiding offence**: the use of some exotic timbers can be offensive as they may come from an unsustainable source.
- **Suitability for the intended market**: beech is a close-grain hardwood that does not splinter, making it ideal for children's toys.
- **Consumer society**: many people feel the need to have the most up-to-date, fashionable furniture. This creates an ever-increasing demand for products, but also leads to an increase in unwanted products.
- **Effect of mass production**: flat-packed furniture can be mass produced using CNC machinery. This reduces the cost of the furniture, making it affordable to a larger audience.
- **Built-in obsolescence**: veneered manufactured boards tend not to be as durable as solid timber. The surfaces may wear through or the edges may chip. This can make this type of furniture obsolete within a few years.

> **Exam tip**
>
> Some questions require you to provide a specific example as part of your answer. Ensure you give an example that is relevant to the material you are describing.

> **Typical mistake**
>
> Some questions may ask you to provide examples to illustrate your answer. Many candidates will miss the fact that the question is asking for *several* examples and not just *one* example.

> **Now test yourself**
>
> TESTED
>
> 1 Name a lightly coloured timber. (1)
> 2 What is meant by the term 'genetic engineering' of timber? (2)
> 3 What is meant by the term 'upcycling'? Give an example of an upcycled timber product. (2)
> 4 What is meant by the term 'planned obsolescence'? Give an example to illustrate your answer. (2)

# 7.4 Forces and stresses

## 7.4.1 Forces and stresses

Table 7.4 Forces and stresses acting on timber

| Type of force | Definition | Example |
|---|---|---|
| Compression | A force that is pushing down on an object | Table leg |
| Tension | A force that is pulling an object apart | Cross brace on a wooden stool |
| Shear | A force that is acting in opposite directions | An unsupported shelf |

- **Natural forces within timber**: the **static force** of heavy branches and the **dynamic force** of the wind blowing against a tree will all contribute to internal stresses affecting a tree trunk. This can affect the timber when it is converted into planks.
- **Pre-stressed construction beams**: a pre-stressed beam is bent during construction to prevent it from bending as much during use.

> **Static force**: a constant force being applied to an object.
>
> **Dynamic force**: a varying force being applied to an object.

# 7.4.2 Reinforcement/stiffening techniques

## Frame structures

- A wooden frame makes an excellent structure when used in the construction industry.
- Triangular roof trusses are used extensively when adding a roof to a house.
- Adding a diagonal strut to a rectangular framework strengthens it significantly.
- Adding a thin plywood panel to a framework increases its rigidity.

**Figure 7.2 A roof truss**

## Fabrication, assembly and construction

The use of wood joints, nails, screws and adhesives can all add strength to a wooden structure. For more information on these techniques see Section 7.7.3 Fabricating and constructing on page 148.

## Lamination

Lamination involves the gluing of several wood veneers over a mould to produce a very strong form. For more information see Section 7.7.3 Fabricating and constructing on page 148.

## Braces and tie bars

- A **brace** is used across a framework to prevent the structure from bowing inwards.
- A **tie bar** is used across a framework to prevent it bowing or distorting outwards.

> **Brace**: prevents a structure bowing inwards.
>
> **Tie bar**: prevents a structure bowing outwards.

## Embedding composite materials

Glass reinforced plastic (GRP), carbon fibre reinforced polymer (CFRP) and natural fibres can all be embedded within a laminated wooden structure to add strength.

> **Exam tip**
>
> When describing a process, it is useful to use subheadings and bullet points, numbers or letters to separate each part of the process.

> **Typical mistake**
>
> When a candidate forgets part of a process, they often feel that they are unable to go back and add extra detail. You should add the extra detail at the end of your answer, but don't forget to clearly show at which stage in the process you would like it to be read.

## Now test yourself

**TESTED**

1. Give an example of a wooden product in compression. (1)
2. Give an example of a wooden product in shear. (1)
3. Describe one method of strengthening a rectangular wooden frame. (3)

# 7.5 Stock forms, types and sizes

It is cost effective to produce designs that incorporate stock forms, as less cutting and shaping will be required.

## 7.5.1 Stock forms/types

**Table 7.5 Stock forms of timber**

| Stock form | Description |
|---|---|
| Regular sections | Natural timber is available in a variety of regular square and rectangular cross-sections |
| Mouldings | **Mouldings** are available in a variety of decorative cross-sections |
| Dowels | **Dowelling** is a circular rod of close-grained hardwood, such as ramin<br><br>Manufactured dowels are prepared ready for use in a dowel joint |
| Sheets | Manufactured timber is sold in sheets that are available in large sizes of varying thicknesses<br><br>They have a smooth surface and have greater stability than natural timber |

**Figure 7.3** Regular sections

**Figure 7.4** Mouldings

**Figure 7.5** Dowels

**Figure 7.6** Sheets

**Mouldings**: long thin strips of shaped wood.

**Dowelling**: long thin circular rods of hardwood.

## 7.5.2 Sizes

- **Planed all round (PAR)** refers to timber that is planed on all sides but will have rounded edges.
- **Planed square edge (PSE)** refers to timber that is planed on all edges with square edges.
- Timber is sometimes referred to in imperial sizes (feet and inches). Common imperial sizes are 8′ × 4′ (2440 mm × 1220 mm) and 2″ × 1″ (50 mm × 25 mm).
- The cross-sectional area is calculated by multiplying the height by the width of a cross-section.
- The diameter is a measurement taken from one side of a circle to the other.
- Manufactured boards are available in standard sizes of 24400 mm × 1220 mm and 1220 mm × 610 mm. They vary in thickness from 1 mm, used for modelling, to 40 mm, which is used for kitchen worktops.

**PAR**: planed all round.

**PSE**: planed square edge.

**Typical mistake**

When asked to give the full definition of an acronym such as PAR, candidates often get one or more parts wrong.

**Exam tip**

Make sure that you know the popular stock sizes for both regular timber and manufactured boards.

## Now test yourself

TESTED

1  Why is it an advantage for manufacturers to use stock size material? (2)
2  What is the common stock size of a manufactured board? (2)
3  What does the acronym PSE stand for? (1)

# 7.6 Manufacturing to different scales of production

## 7.6.1 Processes that can be used to cut and shape materials

REVISED

### Routing

- Router cutters are available in a variety of different profiles.
- A router can be hand-held or table mounted, or controlled by a CNC 3D router.
- They are used to cut slots, shapes and decorative profiles.

### Sawing

- Circular saws, bandsaws and radial arm saws are industrial saws that are used to cut and prepare timber to size and length.
- For more information on saws see Section 7.7.2 Shaping on page 147.

### Mortising

A **mortising** machine cuts a square hole in timber and is often used as one part of a mortise and tenon joint.

> **Mortising**: cutting a square or rectangular hole in wood.

> **Exam tip**
>
> If you cannot remember the name of a woodworking tool when describing a process, make an accurate sketch of it instead.

### Laminating

The lamination process is described in Section 7.7.3 Fabricating and constructing on page 148.

## 7.6.2 Scales of production

REVISED

Table 7.6 Scales of production for timber

| Scale of production | Advantages | Disadvantages | Examples |
|---|---|---|---|
| One-off | A high-quality, unique product is produced using high-quality materials | It usually involves a specialised process requiring a highly skilled workforce<br><br>Expensive | Bespoke piece of furniture<br><br>Built-in wardrobes to fit an individual house |
| Batch | A limited number of identical products are produced<br><br>The cost of the product is reduced as materials can be bought in bulk and processing times are reduced | Initial set-up cost involving the purchase of machinery and manufacturing aids | A set of identical chairs for a high-quality dining table<br><br>Frames for sofas<br><br>Wooden bedframes |

**Table 7.6** Scales of production for timber (continued)

| Scale of production | Advantages | Disadvantages | Examples |
|---|---|---|---|
| Mass | Large quantities of identical products are produced<br><br>The cost is further reduced as computer-aided manufacture (CAM) can be utilised | High initial set-up cost as there is intensive use of machines<br><br>Products lose their individuality | Flat-pack furniture such as TV units, kitchen cupboards, kitchen worktops and bookcases |
| Continuous | The unit cost of a product is significantly reduced as materials can be purchased in vast quantities<br><br>A very high demand can be satisfied | Initial set-up costs are very high, often requiring a dedicated factory to be built<br><br>There is likely to be extensive use of robotics and CAM | Sheets of manufactured timber such as plywood and MDF |

> **Typical mistake**
>
> Candidates often confuse batch production and mass production. Make sure you can give an example of each to help qualify your answer.

## 7.6.3 Techniques for quantity production  `REVISED ☐`

### Marking out

- Measurements can be transferred on to timber using a ruler and a pencil or marking knife.
- All measurements should be taken from an accurate starting point known as a datum line.
- The face edge and face side are planed flat surfaces that are used for marking out.
- A try square will mark a 90-degree line to an edge.
- A marking gauge will produce a line parallel to an edge.
- A mortise gauge will produce a double line parallel to an edge.
- A mitre square will produce a 45-degree line to an edge.
- A sliding bevel can be set to produce any angle.

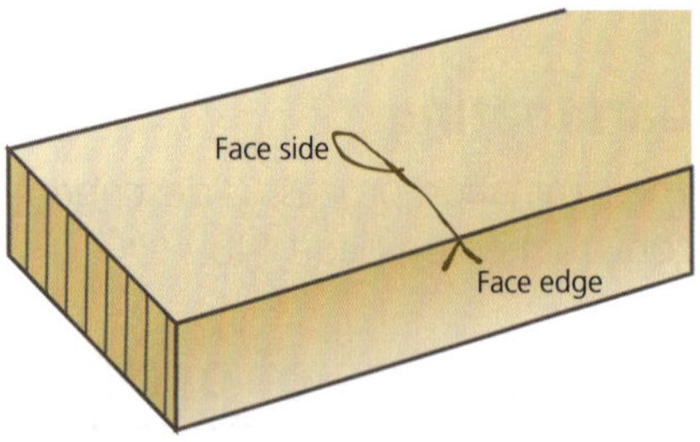

### Jigs

- A **jig** is a specially made device used to assist production by speeding up the process and improving both accuracy and consistency.
- Jigs can be used when sawing or drilling timber.

**Figure 7.7** Face side and face edge marks

> **Jig**: a 3D device used as an aid to production.

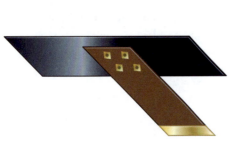

**Figure 7.8** A mitre square

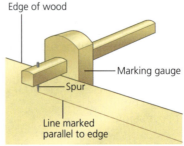

**Figure 7.9** A marking gauge

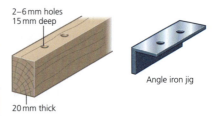

**Figure 7.10** A drilling jig

## Fixtures

Fixtures hold different parts of a timber product together while they are being glued or assembled.

## Templates

- Templates speed up production when several identical components need to be cut.
- Templates can be made from paper, card or more resistant materials such as MDF or plywood.
- They can either be drawn around or used as a profile to be cut around.

## Patterns

- A **pattern** is a 3D version of a template.
- Wooden patterns are used in the process of casting aluminium.

> **Pattern**: a 3D wooden template.

## Sub-assemblies

- A sub-assembly is a completed sub-section that forms part of the main product.
- When making a batch of wooden tables, a sub-assembly of frames would be made before the tops are fitted.

## Computer-aided manufacture

- All CAM begins with a computer-aided design (CAD) program.
- Laser cutters can be used to cut and etch thin sections of timber.
- Thicker sections of timber can be cut and shaped with a 3D router.

## Quality control

- Quality control of timber-based products involves the checking and testing of parts to ensure that they are accurately produced and free from errors.
- Timber should be checked for defects such as splits, knots, bowing, cupping and twisting.

## Working within tolerances

- A tolerance is given as a maximum and minimum size that a component can be.
- A chair leg that is to be 600 mm long could be given a tolerance of 0.5 mm. This would mean it could be any size between 599.5 mm and 600.5 mm and still be acceptable.

## Efficient cutting to minimise waste

- When producing several timber parts from a sheet of plywood it is important to arrange the pieces in a way that will minimise waste. This is known as **tessellation**.
- You must always allow for sawing when arranging/marking out on timber.

> **Tessellation**: the arrangement of components to minimise waste.

# 7.7 Specialist techniques, tools, equipment and processes

## 7.7.1 Tools and equipment

REVISED

- **Hand tools**: there are many hand tools that are specific to working with timber. The advantage of using hand tools is that they are relatively inexpensive; however, they do require a lot of effort and a great deal of skill to use accurately.
- **Machinery**: powered machines take the physical effort out of working with timber; however, they can be expensive to buy.
- **Digital design and manufacture**: computer-aided machines such as laser cutters and 3D routers use CAD to manufacture products quickly, accurately and consistently.

## 7.7.2 Shaping

REVISED

### Drilling

Drills make round holes in timber. Drills can either be cordless or they can be set up in a pillar drill for greater accuracy.

**Table 7.7** Types of drill bits

| Drill bit | Use | Advantages | Disadvantages |
|---|---|---|---|
| Twist bit | General purpose drilling of circular holes | Relatively inexpensive<br><br>Available in a range of sizes from 1 mm to 13 mm | They can become blocked and begin to burn the timber |
| Flat bit | Drilling larger holes | Available in larger sizes<br><br>Quickly removes waste material | Can be difficult to use, creates a lot of torque |
| Forstner bit | Drilling accurate flat-bottom holes | Quick and accurate | More expensive than a flat bit |
| Auger bit | Drilling deep holes in timber | Drills much deeper holes than other drills | Expensive to buy and must be operated slowly, usually in a hand-powered brace |
| Hole saw | Drilling large circular holes | Quickly drills a hole around the perimeter leaving the waste as a solid plug | Often leaves a rough finish, cannot produce a blind hole |

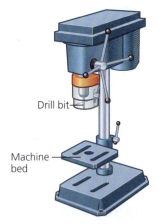

Drill bit

Machine bed

**Figure 7.11** A pillar drill

Centre bit

Jennings pattern auger bit

Forstner bit

Expansive bit (adjustable)

**Figure 7.12** A selection of drill bits

# Cutting

Timber can be cut using a variety of different types of saw.

**Table 7.8 Types of saw**

| Saw | Use | Advantages | Disadvantages |
|---|---|---|---|
| Hand saw | General purpose saw for cutting larger boards and planks | Cuts timber quickly and can deal with large sections/sheets of timber | Can leave a rough finish, requires a lot of effort, not as accurate as other saws |
| Tenon saw | General purpose saw for cutting smaller sections of timber | Can be used with greater accuracy due to its stiffened blade | Cannot cut deep sections of timber |
| Coping saw | Used for cutting curves in timber | Inexpensive, blade can be replaced quickly, can cut very tight corners | Not very accurate, can leave a rough edge, blade can snap easily |
| Scroll saw | Used for cutting shapes in thin sections of timber | Takes the effort out of sawing; can cut fine, intricate shapes | Expensive to buy, the blades wear out/break easily, cannot cut deep sections of timber |
| Jig saw | Used for cutting larger sections of timber | Takes out the effort of sawing, can cut around corners, blade can be replaced quickly | Expensive to buy, blade can wander, leaves a rough edge |

# Planing

- A plane can be used to smooth the surfaces or edge of a piece of timber.
- Special planes, such as spokeshaves, can smooth inside and outside curves.

# Chiselling

- Chisels are used to shape timber by slicing away layers.
- Chisels can be used to remove waste when cutting wood joints.

# Turning

- Timber can be made into a bowl shape by screwing it to a face plate, mounting it on a wood lathe and shaping it using wood **turning** tools.
- Timber can be made into a long cylindrical shape by mounting it between centres on a wood lathe and working it with wood turning tools.

**Turning**: a method of making a wooden blank into a cylinder.

# Abrading

- Timber can be made smooth by sanding it with glass paper.
- Glass paper is typically graded from P80 (a rough paper) to P240 (a smooth paper).
- Sanding is usually done in the same direction as the grain.
- Mechanical sanders, such as palm sanders, disc sanders and linishers, speed up the process.

**Abrading**: smoothing the surface of a material with abrasive papers.

# Carving

Timber can be carved into 3D shapes using specialist carving chisels.

## Rasps and surforms

Rasps and surforms are similar to files but specifically designed for use with wood.

# 7.7.3 Fabricating and constructing

REVISED

## Lamination

Lamination involves the gluing together of thin sheets (veneers) of timber into a curved shape. The **laminating** process is as follows:

1 Firstly, a former is produced.
2 Veneers of wood have glue applied to them.
3 The veneers are layered over the former.
4 The veneers are then held in place by clamps until dry.
5 The laminated timber is then removed and trimmed.

## Veneering

- Exotic hardwood veneers can be applied to manufactured timbers to enhance their appearance. The process of **veneering** is as follows:
  1 A layer of glue is applied to a manufactured board, then a decorative veneer is placed on top.
  2 The assembled board is put into a bag press, sealed and left until the glue has dried.
  3 The board is then removed from the bag press, trimmed and sanded.
- A bag press is a heavy-duty vacuum bag that is used to apply pressure on a product that is being glued.

## Use of screws

Screws are a non-permanent method of fastening pieces of timber together. The process of screwing is:

1 A pilot hole is drilled through both the top layer and the base layer of timber.
2 The top layer is drilled again with a clearance hole.
3 A countersink drill bit is used to recess the top layer.
4 The screw can now be screwed into the joint with a screwdriver. See Figure 7.16.

## Nailing

Nails provide a temporary method of joining sections of timber together. When used with an adhesive, the nailed joint becomes permanent.

- Nails are hammered into the timber.
- Round wire nails have a large flat head to prevent them pulling through thin materials.

**Figure 7.13 Using a surform**

> **Laminating (timber)**: gluing veneers of wood together.

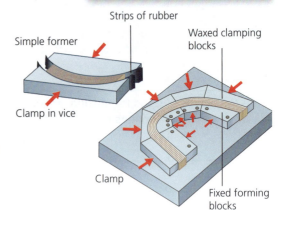

**Figure 7.14 The process of lamination**

> **Veneering**: a thin sheet of wood is glued to the surface of a board to enhance its appearance.

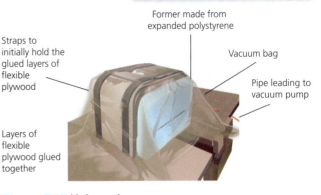

**Figure 7.15 Using a bag press**

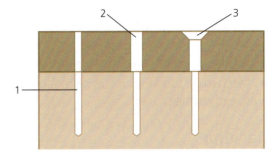

**Figure 7.16 The process of screwing timber**

- Oval nails are shaped so that they are less likely to split the grain.
- Panel pins are small and delicate. They are used to join thin sections of timber and board to thicker pieces of timber.

## Adhesives

### PVA

- Polyvinyl adhesive (PVA) is the most common adhesive used for gluing wood to wood.
- It can be used straight from the tube, is very strong and dries clear.
- It is a water-based adhesive with no specific health and safety issues.
- Its main disadvantage is that it takes a relatively long time to dry.

### Contact adhesive

- Contact adhesive is a medium-strength glue that sticks on contact.
- It is applied in a thin layer to both surfaces and left to dry. The two surfaces are then pressed together and an instant joint is made.
- It is a solvent-based adhesive that is flammable, toxic and an irritant.

> **Exam tip**
>
> When answering a question about using adhesives, make sure that you include details that relate to the preparation of the wood and the adhesive.

## Jointing

**Table 7.9 Wood joints**

| Joint | Image | Use | Advantages | Disadvantages |
|-------|-------|-----|------------|---------------|
| Butt | | Simple frame/box constructions | The simplest of joints to produce | A weak joint |
| Dowel | | Frame/box constructions | A concealed joint that provides greater strength than a butt joint | Usually requires a jig to assist in lining up the holes |
| Lap | | Frame/box constructions | The recess adds strength and provides greater accuracy | Limited strength and requires more time and skill to produce |

**Table 7.9** Wood joints (continued)

| Joint | Image | Use | Advantages | Disadvantages |
|---|---|---|---|---|
| Housing | | To hold a shelf or a divider | The groove adds strength and accuracy | Requires time, skill and a knowledge of several tools |
| Mitre | | Used on picture frames | An aesthetically pleasing joint | Can be difficult to produce without specialist equipment |
| Mortise and tenon | Mortise   Tenon | To construct table underframes and chairs | A strong joint | Requires time, skill and a knowledge of several tools |
| Dovetail | | A traditional method of constructing drawers | A very strong joint | Requires time, a high level of skill and a knowledge of several tools |

> **Typical mistake**
>
> Many candidates get confused with the terms permanent and non-permanent when describing joints. Make sure you can categorise them.

## Wastage

- Wastage is the name given to the process of removing waste material to produce a joint or a shape in wood.
- Wasting processes for wood include sawing, planing, chiselling, drilling, turning and sanding.

## Addition

- Addition is the name given to the process of adding material together to produce a joint or a shape.
- Addition processes for wood include assembling with wood joints, nailing, screwing and gluing.

## Knock-down fittings

- **Knock-down fittings** are used extensively by manufacturers of flat-pack furniture.
- They allow the furniture to be transported home in a flat box and assembled by the customer.
- Knock-down fittings include the corner block, cam lock and scan fitting.

> **Knock-down fittings**: commercially produced mechanical joints used for flat-pack furniture.

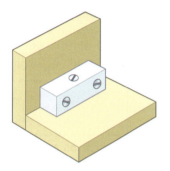

**Figure 7.17 Corner block**

Screwdriver slot

Threaded rod

**Figure 7.18 Cam lock**

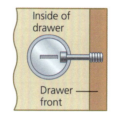

Inside of drawer

Drawer front

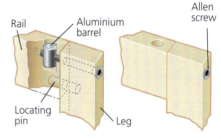

Rail

Aluminium barrel

Allen screw

Locating pin

Leg

**Figure 7.19 Scan fitting**

## Hinges

- Hinges are fitted to a frame to allow doors and lids to open and close.
- The most popular types of hinge are the butt hinge and flush hinge.

**Figure 7.20 Brass butt hinge**

**Figure 7.21 Flush hinges**

## Ironmongery

- **Ironmongery** is the general term given to metal parts that are added to a timber construction.
- Examples include hinges, handles, hooks, knobs, draw runners and locks.

> **Ironmongery**: metal fittings that are attached to wooden products, such as handles and hinges.

### Now test yourself

TESTED ☐

1. Name a hand-powered saw that will cut around corners. (1)
2. Give the name of a mechanical sanding machine. (1)
3. Explain the advantages of using PVA glue. (3)
4. Describe the process of laminating timber. (4)
5. Describe the process of producing a countersunk screwed joint. (4)
6. Explain the advantages to the customer of using knock-down fittings to produce furniture. (3)

# 7.8 Surface treatments and finishes

## 7.8.1 Surface finishes and treatments

**Table 7.10** Surface finishes and treatments for timber

| Surface treatment | Use | Description/application | Advantages | Disadvantages |
|---|---|---|---|---|
| Paint | Used extensively on indoor and outdoor furniture | A coloured liquid applied by brush, rag or spraying | Available in a wide range of colours in a **matt**, **satin** or **gloss** finish<br><br>Easy to apply<br><br>Gives good protection | Conceals the natural grain of the wood |
| Stain | Used extensively on fencing and sheds | A coloured liquid that soaks into the surface of the timber<br><br>Applied by brush, rag or spraying | Changes the colour of the timber while retaining the grain and natural look of the wood<br><br>Outdoor versions give a high level of protection against the weather | Only gives a matt or satin finish |
| Varnish | Used on indoor furniture and wooden boats | Gives a clear, matt, satin or gloss coating to the timber<br><br>Can be coloured with a stain<br><br>Applied by brush or spraying | Provides a hardwearing finish to the timber while offering good protection | Can be affected by sunlight<br><br>Easily scratched or chipped |
| Wax | Used on indoor furniture | Adds a shine to natural, stained or varnished timber<br><br>Usually applied with a cloth | Easy to apply<br><br>Gives a smooth shiny finish and smells nice | Provides very little protection<br><br>Needs to be reapplied regularly |
| Oil | Used on indoor and outdoor furniture | A clear liquid that soaks into the surface of the timber<br><br>Applied by brush, rag or spraying | Easy to apply<br><br>Danish oil gives a high-quality shine<br><br>Teak oil gives a good level of protection | Needs to be reapplied regularly |
| Shellac | Used on high-quality indoor furniture | A natural resin that is applied using a highly specialised technique known as French polishing | Gives a very high-quality finish | Requires a high level of skill to apply |
| Veneering | Used on luxury car dashboards | A thin layer of wood is glued to a manufactured board | Gives the appearance of a high-quality, expensive timber | Requires a high level of skill to apply<br><br>It still requires finishing |

Matt: a dull finish with no shine.

Satin: a finish with some shine.

Gloss: a very shiny finish.

**Exam tip**

When describing a finishing process, it is very important to include details relating to the preparation of the timber, such as sanding, filling and cleaning.

**Typical mistake**

When answering a question about finishing wood, make sure that you have understood if the finish is for an outdoor or indoor application. For example, French polishing is a very high-quality indoor finish that would not be a suitable for a garden fence.

## Now test yourself

TESTED

1   Name a clear finish used on wooden boats.                                   (1)
2   Name a suitable finish for a brightly coloured children's wooden toy.       (1)
3   What type of wooden product would use shellac?                             (1)
4   Explain the advantages of staining timber.                                 (2)
5   Explain the disadvantages of a wax finish.                                 (2)

## Exam practice

1   Give a property of balsa that makes it suitable as a modelling material.   (1)
2   Give a property of larch that makes it suitable for outdoor cladding.      (1)
3   Describe the features of a deciduous tree.                                 (3)
4   Describe the advantages of using manufactured boards.                      (3)
5   Describe the process of veneering.                                         (4)
6   What is mean by the term 'seasoning'?                                      (2)
7   What are the advantages and disadvantages of using contact adhesive?       (3)
8   What is meant by the term 'French polishing'?                              (3)
9   What are the advantages to the manufacturer of flat-pack furniture?        (5)
10  Discuss the issues concerned with using hardwoods from the Amazon rainforest.  (5)

# Success in the examination

## When will the exam be completed?

The exam will be taken in the summer exam period in the final year of study. This is usually May or June.

## How long will I have?

You will have two hours to complete the exam. It is advisable to practise answering sample questions and completing past papers in the allotted time.

## What type of questions will appear?

### Section A: Core content

This will contain a mixture of different question styles including short answer, multiple choice, graphical, calculations and extended-response questions.

You are expected to answer all questions.

### Section B: Material categories from your chosen area

The questions in this section will assess your in-depth knowledge and understanding of your chosen material category – one from either metals, papers and boards, polymers, systems, textiles or timbers. This section will consist of different question styles including short answer, calculations, diagrams and extended response questions.

## Tips on preparing for the exam

- During revision you may encounter topics that you are unsure about or simply do not understand. Make a note and ask your teacher to explain these topics to you.
- Plan your revision timetable well in advance of the exam. Test yourself after each topic is revised.
- Take regular breaks during revision. This will help you to stay focused and retain more information.
- Make good use of past papers, online materials and this revision guide to help practise exam questions.
- Revision cards are a useful method of summarising key facts. Test yourself or ask a family member.
- Work with other students – test each other.

# Sample examination questions

## Section A

### Sample question: Core content

**1 a** The pictures in the table below show four energy sources. Complete the table by naming the energy source and stating whether it is renewable or non-renewable. Part of the table has been done for you.　(6)

| Picture of energy source | Energy source | Renewable or non-renewable? |
|---|---|---|
| | Coal | |
| | | |
| | | Renewable |
| | | |

**b** Biomass schemes burn biofuels to provide energy. Some schemes are considered carbon neutral. Explain what this term means and how it is achieved.　(4)

**Candidate response**

1 a

| Picture of energy source | Energy source | Renewable or non-renewable? |
|---|---|---|
| | Coal | Non-renewable |
| | Wind | Renewable |
| | Solar panel | Renewable |
| | Oil | Non-renewable |

b   Burning of biomass fuels to generate electricity releases carbon dioxide into the atmosphere. In some schemes trees are replanted, which absorb carbon dioxide, making the system carbon neutral.

**Assessment comment**

1 a   One mark for each correct answer, in this case five marks. The answer 'solar panel' is incorrect. The source of energy is the sunlight. The candidate has named the equipment that harnesses the energy, not the source.

b   Burning of biomass fuels to generate electricity releases carbon dioxide into the atmosphere (1). In some schemes trees are replanted, which absorb carbon dioxide (1), making the system carbon neutral.

The points made in the answer, although correct, are not fully explained. A description of what biomass fuels are would further explain the process, for example wood chips, waste from the timber industry. This would lead to the next point and the relevance of planting trees. The trees are replacing ones that have previously been felled but, more importantly, their ability to absorb carbon dioxide as they grow. Carbon emissions are offset – the burning is offset by replanting trees. Overall there is no net release of carbon dioxide into the atmosphere (carbon neutral), although it takes significantly longer for trees to grow than to burn them.

# Section B

## Sample question: Metals

**2** The picture below shows a steel motorway crash barrier that has been galvanised.

    **a** Explain the advantages of galvanising steel. (4)

    **b** Describe the process of galvanising. (4)

### Candidate response

**2 a** Galvanising will protect the steel barrier from rusting by forming a protective layer between the steel, rainwater and air. This will increase the durability of the barrier, meaning that it will last for many years.

    **b** The steel is immersed into a bath of molten zinc. This can be repeated several times to increase the thickness of the coating.

### Assessment comment

**2 a** The candidate has given a full answer. When asked to explain something, you must make a comment and then justify it. In this case the candidate has correctly identified that galvanising will prevent the steel from rusting and has qualified this by adding that it will provide a protective layer between the steel and the rainwater and air. They have made a second point that this will improve the durability of the barrier and stated that this means it will then last for many years.

    **b** The candidate has produced a partial answer. They clearly understand the process of galvanising but have failed to add any details referring to material preparation. They could have also added details such as what temperature the zinc should be.

## Sample question: Papers and boards

**3 a** Describe **two** properties of corrugated card that make it suitable for use as packaging for a takeaway pizza. (4)

    **b** State **two** uses for solid white board. (2)

    **c** Embossing is often used on paper and board products such as business cards. Describe the embossing process. (4)

### Candidate response

**3 a** Good heat insulation properties so the pizza stays warm. Rigid due to the internal structure.

    **b** Picture framing, mounting and presenting of artwork and photographs.

    **c**
- A former of the design is made.
- The former is heated.
- The card is inserted into the former and pressure is applied forcing the card into shape.
- The dies are removed and the design is embossed on the card.

**3  a**  One mark for stating the property and one mark for explaining why this makes it suitable.
In this case the candidate would score three marks. The candidate scores one mark for heat insulating properties and one mark for explaining why this makes it suitable (it stops the pizza going cold). One mark for rigid due to the structure, but no marks for the explanation (it stops the pizza getting squashed).

   **b**  One mark for each correct response. In this case the candidate has given two correct responses.

   **c**  One mark for the use of a former/mould. One mark for male and female dies or parts. One mark for the use of heat. One mark for the use of pressure forcing the card into shape.
This candidate would score three marks. They have described all the stages but not mentioned the need for male and female parts of the former to form the required shape.

## Sample question: Polymers

**4**  The picture below shows mobile phone cases that have been made by injection moulding.

   **a**  Name a suitable polymer for the phone cases.                                                              (1)

   **b**  Use notes and sketches to describe the process of injection moulding.                                      (6)

**Candidate response**

**4  a**  Plastic

   **b**  Plastic pellets are emptied into the hopper. They fall into the heating chamber and are carried along by a large screw. As they move along the heat changes the pellets into liquid plastic. When it reaches the end of the heating chamber it is injected into a mould. The mould is then cooled, and the product is ejected out.

**Assessment comment**

**4  a**  The candidate has used the generic term 'plastic'. This is not enough for a mark. The candidate should have named a specific thermoforming polymer. Acrylonitrile butadiene styrene is the correct answer; the abbreviation, ABS, would also be acceptable.

   **b**  The candidate has produced a detailed written answer to the question. Although some technical terms are missing, for example 'large screw' instead of 'Archimedean screw', there would be sufficient information for full marks to be awarded if the candidate had also produced a sketch. The lack of a sketch, which is specifically asked for, would result in the loss of a mark.

## Sample question: Systems

**5** Explain **two** advantages of through-hole circuit construction over surface mount technology. (4)

5 • Advantage 1: Through-hole construction results in stronger soldered joints. This means that they are less likely to fail over time.
  • Advantage 2: The components and PCB are much larger, which makes it easier to locate and repair faults within the circuit.

**Assessment comment**

This candidate would get full marks. The question asks you to explain two advantages, so the answer must go beyond two simple statements. You would usually receive one mark for each valid advantage and a further mark for the explanation of each.

Common errors in this type of question include not explaining your answer sufficiently or only writing about a single advantage instead of the two requested. You are not being asked to write about disadvantages, so only add them if the question specifically asks for them.

## Sample question: Textiles

**6** Fashion and textile designers use a number of techniques and processes to improve the structural integrity of products. The fabric shown in the picture below is ripstop nylon.

**a** Give **one** reason why nylon is a suitable fabric for a surf kite, as shown in the picture below. (1)

**b** Evaluate the use of a ripstop nylon – a rib weave fabric – for windsurfing kites. (6)

**Candidate response**

6 a One of the properties of nylon is strength, so it will withstand pressure when flying.
  b The fabric needed for a surf kite would need to be strong and stable. A rib weave is a variation on a plain weave but is much stronger. This weave would withstand the pressure and force that the kite will come under when in use and, therefore, is a good choice for a surf kite. Rib weave is abrasion and tear resistant.

**6 a** An acceptable answer for one mark. Note that nylon also has good elasticity, allowing some 'give' in the fibres when under tension; this would prevent rips or tearing.

**b** In responses to *evaluate* questions, evidence of appraisal or making a judgement needs to be clear. There is some evidence of this within this response. This answer would gain three marks (the lower mark in level 2).

The candidate has a fair understanding of the essential properties a fabric should have for a surf kite and has some understanding of the weave being different to a plain weave. The points made are not developed or substantiated, however.

To gain further credit the candidate should explain the technical difference in a rib weave when compared to a plain weave, that is, the additional thicker thread intermittently placed that strengthens the weave and makes it abrasion and tear resistant. This is an important point that has been omitted. Further explanation of how this impacts the functionality of the surf kite would also be required.

## Sample question: Timbers

**7** The picture shown below is of a child's wooden toy racing car.

**a** Name a specific timber for the child's wooden toy racing car. (1)

**b** Describe and explain the properties of the timber named in (a) that make it suitable for the child's wooden toy racing car. (4)

**c** Describe how you would make the body of the car in a batch of ten. (6)

**7 a** Beach

**b** Beach is a tough, durable, close-grained hardwood that will take the punishment inflicted by a child playing with it. It does not splinter, therefore it is safe for children to play with, even if they decide to chew it.

**c** I would first mark out the shape of the car on the wood. Then I would clamp down the wood with a G clamp, protecting the surface with a scrap piece of wood. Then I would use a tenon saw to cut the outside shape. I would then use a mixture of files, surforms and glasspaper to smooth the shape. Finally, I would drill the holes for the axles with a twist bit held in a pillar drill.

**7 a** Although the candidate has misspelled beech, they would be still be awarded the mark.

**b** The candidate has produced a full and complete answer, clearly giving several correct properties of beech that make it a suitable timber and clearly linking each property to its application.

**c** The candidate has given a good description of how you would produce the body of the car as a one-off. They have lost marks for failing to address how it would be made in a batch of ten. They could have included details such as drawing around a template instead of marking out. They could also have suggested using powered machinery, such as a scroll or bandsaw, and the shape could have been made smooth by using powered sanders, such as a disc or palm sander.